# THE
# MUSHROOM
# BASKET

# *Frontispiece*

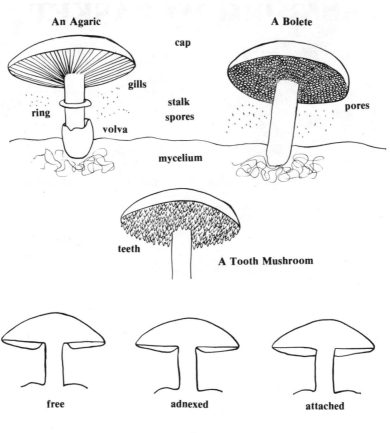

**An Agaric**

cap

gills

**A Bolete**

ring

stalk
spores

pores

volva

mycelium

teeth

**A Tooth Mushroom**

free

adnexed

attached

**Gills**

# THE
# MUSHROOM BASKET

A Gourmet Introduction to the Best Common
Wild Mushrooms of the Southern Rocky Mountains,
With Applications Throughout the Northern
Hemisphere, and Tidbits of Mushroom Lore from
Europe, Russia and China

**Andrew L. March**
**Kathryn G. March**

**MERIDIAN HILL PUBLICATIONS**

Please: mushroom gatherers must pay attention and must study, because a few mushrooms are deadly.

With due care, you will find the woods and fields safer in respect to all kinds of food poisoning than hamburger stands and supermarkets.

This book is as accurate and foolproof as the authors can make it. But please use it with the understanding that they refuse legal responsibility for mishaps.

*for*

Patricia Pendleton Amazalorso

# Acknowledgements

Many persons, mushroomers and non-mushroomers, helped and encouraged us with this book. Especially generous with their time and wisdom were:

Linnea Gillman, Denver Mycological Society and (formerly) U.S. Forest Service, who examined the whole manuscript and made invaluable comments;

Cheryl Johnson, former editor of the *High Timber Times,* who gave us the benefit of her editorial and mushrooming expertise throughout;

Roger and Julie Sliker, who introduced us to *Boletus edulis* and some of the other best species in the Colorado mountains;

Joan Wheeler, frequent guide and companion, who actually showed us some of her secret mushrooming grounds;

Ying Jianzhe of the Institute of Microbiology, Academia Sinica, Beijing; Mrs. Ying helped greatly in determining which of the mushrooms grow in China, and what their Chinese names are.

We are grateful to Pat Amazalorso, Jan Dalrymple, Ginny and Bob Massey and John Wiegand for their help and support. Marilyn Shaw was good enough to help us locate three missing photographs, and we thank Athalie Barzee, Paul Sandoz and Dennis Thurber for letting us use their slides of *Morchella angusticeps, Agaricus sylvicola* and *Pluteus cervinus,* respectively.

We also express our appreciation to Morris Schertz, Director of Penrose Library at the University of Denver, for granting us access to the Chinese and other collections there; and to Solange Gignac, Librarian at the Denver Botanic Garden, for making us at home in that library.

These people, of course, are not responsible for the book's defects, which are ours alone.

# Forward

In this book NEWCOMERS will find the information they need for quick and confident identification of the handful of species that make up most of the wild mushroom harvest in the southern Rocky Mountains, in fact, in the whole northern hemisphere. They will be steered away from poisonous kinds, and encouraged in safe mushrooming habits.

EXPERTS may enjoy new slants on familiar mushrooms with lore from Europe, Russia and China.

ALL we hope will find pleasure in the pictures and poems, and in the international gourmet recipes that enshrine fullest appreciation of the inimitable qualities of wild mushrooms.

Observers of green plants or birds are accustomed to being able to identify in their manuals any individual they come across. But mushroomers, even experts, must accept finding many mushrooms they cannot put an exact name to. There are no complete guides, even regional.

The mushrooms in this book are those generally most profitable to know. They are common enough so that in season there is almost sure to be something to bring home from any excursion. And they are absolutely TOP-DRAWER things to eat; regardless of rank or riches, or the utmost snobbishness, no king nor emperor could contrive any fancy imported high-flown delicatessen gourmet tidbit that would be superior to the WILD MUSHROOM.

# Contents

### *Boletus edulis*

The midlife couple
and the dancing child
pass over the woven
distances, the pinewood's
scattered scent and shade,
the broad slope of the mountain.

It's not Christmas, not Easter
but an older fiesta
held by the forest
sometimes in August.

Now separate, now together
they watch the speckled ground
and they study each other,

whose paths zigzag and crisscross
whose feet point east and west,
as they stop, pivot, stoop,
straighten, and again
move off with their basket.

# Introduction

## Biology

MUSHROOMS in the ordinary sense of the word are not the whole Kingdom of the Fungi with its molds, rusts, yeasts, and slime molds, but only the larger fleshy species that are conspicuous and look as if they might be good to eat.

The perennial mode of existence of mushrooms is not the familiar cap and stem: these are fruits, and come and go from season to season. But within the earth, wood, manure or other organic material from which the mushroom draws its nourishment lives a mass of fine threads, often white. These are the hyphae; together they are called the mycelium. A mycelium may last years or decades or longer, and is thus the constant self of the mushroom.

When temperature, moisture and other conditions are right, the mycelium forms tiny knobs which poke up and out into the atmosphere as fruiting bodies. When they mature their mission is to release spores which can move away from the parent plant and start a new mycelium. The spores are very numerous and very light, and are dispersed by wind and water and other agents. Only a very small fraction survive to start fresh growths.

The fruiting bodies (see Frontispiece) typically consist of a stalk which elevates a cap high enough to expose its underside to the air. Beneath the cap are gills, pores or teeth that maximize the surface upon which the spores are borne, arranged so that the spores may fall free as they ripen. Gills are vertical strips of flat tissue that radiate from the stalk to the outer edge of the cap. Pores are the mouths of vertical tubes, small to tiny. Teeth are little soft spines or fingers that hang down from under the cap. There are good edible mushrooms of all three kinds. The most poisonous ones are among the gilled.

Some mushrooms have sheets or threads of tissue that connect the stalk with the cap before it spreads open. This veil may leave a ring around the stalk when it tears loose.

Some, especially the dangerous genus *Amanita,* are wholly enclosed in a universal veil when very young, in the button stage. When it ruptures as the mushroom expands, it leaves a volva: a cup, sack, sheath, or ridges or fragments of tissue at the stalk base. It may leave remnants also on the top of the cap.

Morels have less distinction of cap and stalk and bear their spores on a pitted and ridged upper surface. Puffballs carry theirs inside and release them through a hole or tear.

Mushrooms live in many habitats, from desert to tundra, from woods to fields to ploughed ground. But the most fruitful areas are the great forest zones of the low latitudes and--where the mushrooms this book is concerned with are found--of mid-latitude Eurasia and North America. Within their latitudes mushrooms are quite cosmopolitan. Many species grow in the Asian, African and American tropical forests, but with little overlap into temperate regions. Mid and high latitude mushrooms are widely shared across Eurasia and North America all around the north part of the planet. Some common genera, though, are in both zones: there are a few *Russula, Lactarius* and *Amanita* in East Africa, but nothing to approach the proliferation of species in the cooler forests to the north. The Meadow Mushroom is at home in both zones.

Figures on the number of mushroom species--the fleshy fungi that look as if they might be edible--vary because the count is not yet complete and because mycologists differ as to species, species complexes, varieties and geographic races. The total is upwards of 5,000 for North America and around 3,000, overlapping considerably with ours, in Central Europe. It seems that roughly 10 percent--a few hundred--will prove to be valuable edible mushrooms: wholesome, fairly common, tasty and susceptible to safe identification without laboratory study. Something like one percent--a few tens--are habitually gathered and consumed by large numbers of people in these areas, especially Eurasia.

There are undoubtedly more unused but good species in North America, perhaps in western North America alone, than in the old world temperate zones. The old world may know more, popularly, about mushrooms and be taking better advantage of them now, but North America is richer in species.

So perhaps it is not too bold to predict a magnificent future, rich in new kinds of deeply credible and engaging art, science, poetry and folklore, for mushrooms in America.

Mushrooms are aleatory and capricious: here today, gone tomorrow; fruiting one year, absent the next. But in the temperate zone forest areas there is usually something in the way of gatherable mushrooms all through spring, summer and fall when it is wet enough, and even in winter when it is not too cold. In most places the major season is late summer and early to mid autumn. Certain mushrooms may be found during the whole warm season off and on in the southern Rockies--*Suillus* and *Leccinum* for example. Others are concentrated in big fruitings that can yield enough to both eat and preserve: morels and Oyster Mushrooms in spring, King Boletes and then Honey Mushrooms in August and September, Shaggy Manes on into October, for example. Some can even stand frost, most notably the Winter Mushroom which fruits in mild spells through the winter in some places.

## Names

The idea is that regardless of language, region or personal preference a universally agreed upon Latin name for any plant or animal or fungus or other living thing will always refer to a single species--one "kind" of thing, a continuing population of individuals with a stable range of characteristics. The system has a time frame as well. The species, represented by the second of the two Latin words (e.g. *comatus* in *Coprinus comatus*), is considered to descend from a common ancestral group of individuals somewhere in the past. The genus *(Coprinus)* is like cousinhood: all the Coprinuses (*Coprinus comatus, C. micaceus, C. atramentarius,* etc.) descend from one species still farther back. Similarly a group of genera constitute a family, which was a single species still longer ago; and above family are other broader ranks.

The Latin names are less absolute in practice than in theory, but still they are usually a surer guide than common names; in English, mushrooms usually lack good common names anyway. Mushroom classification is in some turmoil and more than one scientific name is current for some species. Mycologists are still wrestling with genera, families and even the wider categories. Thus in some books *Dentinum repandum* is *Hydnum repandum, Flammulina velutipes* is *Collybia velutipes, Agaricus bisporus* is *Psalliota hortensis.* Big old genera are often split up into several new ones. Many species are coming to be regarded as species complexes, groups of hard-to-distinguish species, or alternatively big loose species with a great deal of internal variability. *Boletus edulis, Pleurotus ostreatus* and many other of the commonest and best are such complexes. It is a fair guess that by the end of the century mushroom classifications will no longer be so topsy-turvy, and the species concept will have stabilized.

The Latin names are barbarities to pronounce in English or any modern language. Most sound in-groupy, exclusive and intimidating. The various half-hearted rules result in a plethora of the longest vowels English has: lots of "i" as in iron, "a" as in Avon, "u" as in utility, which ring ill on the ear. But there is no true "correct" standard for these names, and people quite properly say them everywhichway as sounds to them least offensive. For our part we lean against the long vowels and prefer AmanEEta to AmanEYEta *(Amanita)* and LEXinum to LekSINEum *(Leccinum).*

The best rule is to say the Latin names (1) fast, and (2) confidently.

## Studying

Scientific names and the rest of the scientific apparatus, taken in conjunction with the bad reputation mushrooms/toadstools have in the Anglo-American world, discourage many people from gathering mushrooms, and indeed caution is essential. But even the most dedicated lifelong professional specialists do not know every mushroom they come upon, though they can always say something about genus or at least family. For the majority of mushroom lovers it is not an all-or-nothing proposition. It is perfectly possible to know enough about a few mushrooms to pick them with pleasure and confidence, and to go on studying others over the years, adding now one now another to one's repertory, never imagining that until one recognizes every mushroom in the forest, and has tasted all that are not venomous, one has failed. It is not a study for headlong and impetuous persons; most mushroomers seem to be in their 30s or older. Not that the discipline is uniquely hard in itself, but it does exact continuing attention to detail and patience with delay. It punishes carelessness.

The best strategy is to begin with one or two common, good and easy mushrooms, while at the same time memorizing the most dangerous, i.e. those that can be deadly. There is a lot of psychology in the identification and eating. It is important to feel sure in the mind that everything is right. Many a new eater of wild mushrooms is made queasy later by the ghost of a doubt: what if...? Best to take only a bite or two of a new mushroom; then one can say, well, even if it was a mistake, it won't finish me.

Mushrooms are in fact so psychological in many ways that it would be no surprise to see a school of Mushroom Psycho-therapists, among the many other schools: How To Externalize And Conquer Your Fears And Inadequacies Through Controlled And Rational Experimentation With Eating Wild Mushrooms.

Once the benchmarks of the most lethal, and the best and the easiest, are firmly set, mushroomers can go ahead and explore as far into the thousands of others--the good but rare or difficult, the not-so-good but usable, the evil-tasting but not poisonous, the mildly indigestible, the merely beautiful--as time and interest carry. It is a splendid recreation, a scientific experience, a survival skill, a contribution to the household economy, good for the health, and something to show off to friends. It's an excuse for being in the woods, and an unending game and surprise.

It is a good idea to look carefully at all kinds of mushrooms from the first, noticing the details--how gills attach to stalk, colors, surface textures, signs of rings or veils, and so forth. *Negative* identification is as important as positive: without looking over many strange mushrooms, it is hard to be certain about how pronounced a feature must be before it can serve as a distinguishing badge of identity. One must be able to say: I don't know what this is, but it ISN'T a so-and-so because it doesn't have such-and-such characteristics.

Some of the best more comprehensive manuals and field guides are listed in the Bibliography. A good avenue for pursuing the interest is a mushroom club, or mycological society, of which there are many in the United States and other countries now. For those who are so inclined, they offer many opportunities to meet with other mushroom buffs and share in their excursions. We have found great profit in the yearly Mushroom Fair of the Colorado Mycological Society held in August at the Denver Botanic Gardens. Fresh specimens are laid out and identified, and some of the finest mushroom experts in the world are there to discuss problems.

Anyone with the reading and writing habit will probably want to keep a card file of mushrooms, with notes on where and when they're found, cooking ideas and so forth.

## Gathering

Baskets are traditionally preferred. The handle makes them easy to carry, and their rigidity keeps the mushrooms from crumbling into a gritty unrecognizable heap. Actually a couple of baskets, plus a paper bag or two, are handy so that edibles can be kept strictly separate from poisonous or unknown specimens. A knife, perhaps with one sharp and one dull blade, for cutting and for digging up stalks; a magnifying glass; and something to write notes on, are all useful.

Plastic bags are not good because mushrooms have a high water content and deteriorate fast if they cannot breathe.

A gatherer needs to be aware of the kind of tree a mushroom grows under, whether it grows in grass, or on wood or manure; whether in clumps or singly. Samples for study should include several mushrooms of the same species, young and old. It is important to collect the whole stalk including the very bottom in case there is a bulblike swelling or the remains of a volva. The ideal time for gathering is when the outdoors has dried out after a rainy spell. Mushrooms last longer when it's cool.

When a mushroom is definitely recognized as an edible, slice off the bottom of the stalk to check for worm holes. Many of the fly and beetle larvae work upwards, so even if there are little pinhole tunnels at the base, progressive slices farther up may reach virgin territory. This is done immediately on picking because the creatures go right on eating in the basket, so a mushroom good in the field may be spoiled by the time it gets home, and even spoil others in contact with it.

Dirt should be brushed off and bad spots trimmed away before an eating mushroom goes in the basket, and it should be laid with gills (pores, teeth) down because grit that gets in there is nearly impossible to wash out entirely.

Mushrooms must be gathered in clean places. They are very sensitive to their environment, and pick up whatever is there, fair or foul; they are easily contaminated with poisonous metals, fertilizers, pesticides and so forth.

## Poison

Besides being economical and mystical, mushrooming is ineluctably intellectual: mushroomers must keep their wits about them. There is danger, but it is chosen and manageable.

All the old wild tales about how to tell the poisonous one from the edible one, the "mushroom" from the "toadstool," or how to render all harmless to eat -- NOT ONE is true. They are products of societies without laboratories and where information and communication were scarce, and in the end they depended on the fact that MOST mushrooms do not happen to be poisonous, so that if people lucked out they might go on gathering the same mushrooms in the same places year after year, with small odds of sickness or death. Their odds of learning of accidents, unless happening to their neighbors, were not great either.

--Mushrooms are poisonous if gathered when the sun is shining on them. Mushrooms are safe if gathered when the moon is full. Mushrooms growing in apple orchards in bloom are edible. Mushrooms that blacken a silver spoon, turn yellow when sprinkled with salt, or change color when cut or bruised, are poisonous; others are safe. Mushrooms that grow on wood are safe; mushrooms that grow on wood are poisonous. If animals eat a mushroom, it's safe. If mushrooms are boiled with pears they become safe. -- ALL FALSE.

And from China: when boiling mushrooms, put in some candlerush or a silver hairpin, or rice grains, or bits of ginger; poisonous mushrooms will turn them black. Make mushroom soup, and look into it: if you cannot see your reflection, the mushrooms are deadly. -- ALL EQUALLY FALSE.

The only sensible procedure is to know and be on the watch for the few most dangerous mushrooms that might be confused with edibles (pp. 23-26); and to eat, at first, only a few easily identified kinds of good mushrooms.

See also the precautions on pp. 26-27.

### *Suillus*

Slippery Jack on the side of the hill
Fry him in butter and eat your fill.

Slippery Jack and some greasy bread
A jar of homebrew to clear your head.

Slippery Jack standing in a row
Where better mushrooms are supposed to grow.

With Slippery Jack you can't go wrong
But the worms'll get him if you wait too long.

Stick your hand out, pick a mushroom
Plenty more where that one came from.

Forget King Bolete and the rest of the pack
What I want is Slippery Jack.

Slippery Jack has gone to the wars
All he left behind is a jillion spores.

# 2. Guide

## Identification

This Guide is designed to remove from consideration all but a few of the mushroom species likely to be encountered, leaving only a basketful of the best, safest and most plentiful.

A prospective edible mushroom should be examined carefully to see which if any of the Five Basic Groups it belongs to. If it fits one of them, turn to the indicated page and see if it will key out more exactly. Check the pictures for hints and confirmations, but pictures alone are not adequate for identification. Look for each feature mentioned, and conscientiously rule out the look-alikes.

The Keys are made up of pairs of mutually exclusive descriptions, A1/A2, B1/B2 etc. Choose whichever term of a pair fits your mushroom, and follow on to the next indicated letter pair, until no description fits (which will often happen) or until the mushroom is identified. The Keys are intended to eliminate undesirable mushrooms as much as they are to identify good ones.

For anatomy and nomenclature, see the Frontispiece and the Biology section (pp. 12 -14). "Attached", of gills, means that they reach all the way in to the stalk and are broadly connected to it (though they may pull away in older specimens); "free", that they do not reach; "adnexed", that only a small part of the gill comes to the stalk. "Close" means the gills are packed closely together.

Some mushrooms "stain" or "bruise" various colors when they are handled, pinched, cut or broken, and these colors are sometimes important in identifying them.

To make a SPORE PRINT of an Agaric (gilled mushroom), remove the stalk and lay the cap with gills down half on white and half on black paper; if it is dry put a drop or two of water in the middle of the cap. Cover with an inverted bowl, and leave for four to twelve hours. The accumulation of spores fallen from the gills will show spore color as white, brown, etc. Natural prints may often be found beneath mushroom caps as they grow.

A "viscid" cap is sticky or even gelatinous when wet, and when it dries out often has twigs, pine needles, etc. stuck to it.

Mushrooms that grow on wood may also seem to grow on the ground when they are with buried wood. Best avoid them if there is any ambiguity.

The measurements given are typical. Larger and smaller individuals may also be found.

## Dangerous Mushrooms

These are the few mushrooms that it is most essential for gatherers to be conscious of so as to avoid eating them by mistake.

If it weren't for a small group of Amanitas, mushrooming would be a whole different game; this one genus is responsible for the large majority of deaths. They are handsome, common, and large enough to attract attention. In fact most are not even poisonous, but they are hard to tell apart and the only sensible recommendation for any but very advanced mushroom hunters is: Leave Them Alone. And also: Watch For Them, Learn To Recognize Them As *Amanita* (Plate 8 , Figs. 1 and 2).

Beautiful and (some of them) deadly -- always a fascinating combination. Amanitas are gilled mushrooms, growing on the ground, mainly in the forest, with white or nearly white spores and caps 1½ " to 6 " across. The gills are free from the stalk, or only slightly adnexed. They have a "universal veil" that leaves a "volva" at stem base in the form of a cup or sack, ridges, a bulb, or scattered powder in the surrounding soil; they may have warts on top of the cap. Most, especially the poisonous ones, have a ring, but this can be fragile and fall off. Stalk and cap separate easily.

Amanitas are why it is essential always to grub up the whole stalk of a gilled ground mushroom in gathering, and to look at it critically. Wear glasses if necessary. Baby Amanitas are completely enclosed by their universal veil, and can be confused with puffballs, which is why puffballs are to be cut vertically in half to make sure there is no embryonic differentiation of cap, stalk and gills. Young Meadow Mushrooms look similar but their gills are pink not white.

The ill reputation of the genus rests above all on five species. Three -- *Amanita bisporigera, A. virosa* and *A. verna* -- are white,

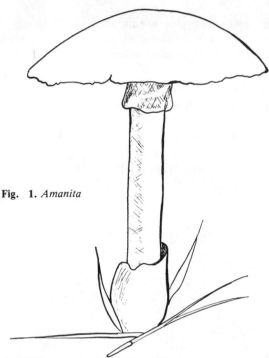

Fig. 1. *Amanita*

and are found in the eastern half of North America (*A. virosa* rarely in the west); *A. ocreata* is white to cream-colored and grows in the southwest. But the ranges are not guaranteed, and all should be watched for wherever one gathers.

These four are called the Destroying Angels. The fifth notorious *Amanita* is *A. phalloides,* a Eurasian species now spreading in North America from the east and west coasts. Its cap is greyish yellow-olive or brownish yellow. All five contain some of the most potent natural poisons that exist.

The gorgeous white-spotted red *Amanita muscaria* or Fly Agaric, and the speckled tawny *A. pantherina* are poisonous-hallucinogenic. The accounts we have read of trips with them, in their American form at least, greatly discourage experimentation. It is the Fly Agaric that provoked so much literature and excitement in recent years by its association with former religious rites and experiences in several parts of the world. But human chemistry is such that it requires no agaric to make the mind fly. Unassisted meditations have a longer and more reliable reach, for people who are serious about what they're doing and not just playing recreational games with their brains.

The reddish-brown-spored gilled genus *Galerina* contains several dangerous mushrooms. *G. autumnalis* grows on wood and can be confused with the Honey Mushroom or possibly with the Winter Mushroom, both of which however have white spores. *G. autumnalis* has attached yellowish gills that darken with maturity to red-brown, a membranous ring at least when young, and a small viscid brown cap, 1 ″ to 2½ ″ wide. *G. marginata* is similar but the cap is not viscid.

*Gyromitra esculenta* is an irregularly shaped brown mushroom with ridged and wrinkled cap (no gills, pores or teeth) 1½ ″ to 4 ″ wide and chambered inside, that might be taken for a morel. But

**Fig. 2.** *Amanita muscaria* **(Fly Agaric)**

morel caps have pits and ridges, and their interior is hollow not chambered. Raw *G. esculenta* is deadly poisonous; cooked just so, it is regarded as edible by some here and abroad, at least in the short run. It seems foolish to try it. See also under morels, p. 31.

It only makes sense for mushroomers to be aware of the nearest poison center, or even to keep on hand a bottle of syrup of ipecac to induce vomiting if a mushroom itself has not already caused it (and unless there is unconsciousness or convulsions) in case expert help is not at hand. With mushrooms that are new and unfamiliar a sample should be saved in the refrigerator to facilitate re-identification in case of difficulties later.

But if the mushroomer has done proper homework, there is no reason for any problems to occur.

## Precautions: Checklist

1. Note where a mushroom grows -- on the ground, on wood, in grass, under what trees.

2. Examine the whole mushroom, including the bottom of the stalk which may be underground.

3. Look with attention at each mushroom picked, once when it is gathered and once again at home, before cooking. Mushrooms do not necessarily grow in homogeneous groups and poisonous species may be mixed among edible ones.

4. For identification, collect several specimens at various stages of development.

5. Keep unknown or poisonous species separate from others in the basket, in paper bags or wax-paper (not plastic) sandwich bags.

6. Don't rely on only one or two characteristics to identify a mushroom. Rings may fall away, colors may fade, attached gills may separate and appear free. Every character must fit.

7. With gilled mushrooms, take a spore print.

8. Gather and study a mushroom a number of times before eating if necessary for confident identification.

9. Eat only a tablespoon of a newly identified mushroom the first time, and wait at least a full day before eating more. Save a fresh specimen in the refrigerator for possible reference later.

10. Eat only one new kind at a time, and with little or no alcohol.

11. Eat only mushrooms that are picked fresh and in good condition. Mushrooms spoil fast and can produce ordinary food poisoning like other foods. Also it's harder to identify old specimens.

12. Cook all mushrooms before eating. Although some may be wholesome raw, many are indigestible or poisonous.

## Five Basic Groups

The idea is to eliminate large categories of mushrooms that are not recommended for beginners or for ordinary gatherers -- persons not concerned with full-dress scientific mycology.

The simplest are first. With the boletes complexity increases, and the agarics are most complicated of all. None given here, though, are rare or especially difficult to identify.

NOT INCLUDED are:

--Very small mushrooms, those whose cap diameters seldom exceed 1½ " in mature individuals.

--Non-fleshy mushrooms: woody, leathery or jelly-like.

--Coral-shaped or fingerlike mushrooms; flat or cup-shaped ones; puffball-like mushrooms with star points that open out; mushrooms without gills that project from wood like shelves.

--Underground mushrooms.

--Mushrooms that grow primarily on manure.

--Mushrooms with cobwebby veils.

I. No real stalk, or no clear division of stalk from top; round to pear-shaped; 1 " to 20 " in diameter; white to tan outside, clear white and firm inside when immature, turning yellowish-brown to purplish-brown and softening with maturity, releasing dustlike powdery brown spores when squeezed.
PUFFBALLS, p. 28

II. Definite distinction of stalk and cap, both hollow, and joined around edge so stalk reaches little or not at all up under the stalk toward apex of cap; brittle; cap more or less conical, pitted with deep depressions that are separated by ridges (not just a wrinkled

cap); mostly in spring.
MORELS, p. 30

III. Distinct cap and stalk, with downward projecting short soft tiny teeth (fingers, spines) on underside of cap.
TOOTH MUSHROOMS, p. 32

IV. Distinct cap and stalk, the undersurface of the cap covered with small vertical tubes.
BOLETES, p. 34

V. Distinct cap and stalk (or stalk may be missing in Oyster Mushroom, *Pleurotus*, growing on wood) with gills on undersurface of cap radiating outward from stalk.
AGARICS, p. 42

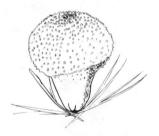

Fig. 3. *Lycoperdon perlatum* (Puffball)

## I. PUFFBALLS

A1 1 " to 2 " in diameter, on wood or ground, releasing spores through small hole on top when mature . . . . . . . . . . . . . . . . . . . . . . . . . . . . . . . . . . . . . . . . . . . . . . . . . . . . . . B
A2 Larger, on ground, outer covering breaking irregularly to release spores . . . . .
. . . . . . . . . . . . . . . . . . . . . . . . . . . . . . . . . . . . . . . . . . . . . . . *Calvatia* and *Calbovista*.

B1 White with minute conical spines that crumble away leaving marks, on ground with grass or trees . . . . . . . . . . . . . . . . . . . . . . . . . . *Lycoperdon perlatum*.
B2 Pale tan to brown, smooth to slightly roughened, on dead wood often in clusters . . . . . . . . . . . . . . . . . . . . . . . . . . . . . . . . . . . . . . . *Lycoperdon pyriforme*.

(Plate 1 , Fig. 3.) Edible puffballs are round to pear-shaped mushrooms without distinct stalks, 1 " to 20 " in diameter. Inside is CLEAR WHITE when young, homogeneous, firm but not hard, darkening and softening with age to a mass of yellowish-brown to purplish-brown powdery spores mingled with cottony dry threads and puffing out spores when pressed. The mature outer covering is more or less papery and fragile. Species of *Calvatia* and *Calbovista* are hard to tell apart.

Puffballs grow in many environments and have a long season through summer and fall. They are edible and delicious so long as the inside remains perfectly white and firm with no trace of yellow or other coloring. If eaten when overripe they are bitter and cause gastrointestinal upsets.

**Look-alikes.** Each puffball should be sliced in half vertically and examined carefully to be sure it is not the unopened button of an *Amanita*--if it is it will show differentiation into a tiny cap, gills and stalk. Puffballs will be homogeneous.

Stinkhorns when young resemble puffballs but are slimy or jellylike inside.

Hard-skinned puffballs whose insides are purple *(Scleroderma)* should not be eaten.

**Cooking.** Puffballs are not at all fibrous and almost literally melt in the mouth like butter. Slice 1/16 " thick, season with salt and basil or oregano, saute till crisp and golden. They are good sprinkled like bacon over cooked vegetables.

Puffballs dry and freeze well. They should be defrosted before sauteing.

**Names.** Besides "pullball" English has a nice history of expressions: fist, puff-fist or puff-foist, puck-fist, and woolfes fistes. The last is explained by Gerard's 1597 *Herbal* as "lupi crepitus," i.e., wolf fart; and "fist" in the other words means the same from other sources. Also devil's snuffbox, fuzz or fuss, fuzzball, etc.; bunt and frog-cheese, for a young larger puffball. For once English matches the continental European languages in its list of common names for a mushroom.

Other languages show the same reactions and preoccupations: Spanish pedo de lobo = wolf-fart, as does the genus name Lycoperdon, from Greek roots. Spanish also says bejín. French: vesse-de-loup, ditto; Italian vescia, ditto. German Stäubling "dust-

Fig. 4. *Morchella* (morel)

mushroom'', but also Bubenfist, Nonnefurzli, etc., like the English words. The idea, of course, comes from the mature not the edible stage.

In Russian the big ones are golovach, the small ones dozhdevik (''rain-mushroom''). Chinese: mabo (''horse turd''), chenjun (''dust mushroom''), wangwen huibao (''reticulated packet of dust'' = *Lycoperdon perlatum*).

**Miscellany.** The dusty spores of mature puffballs are a styptic and have been so used in many lands. Nosebleed can be stopped by puffing the dust up the nostril, but care should be taken not to inhale it. *Lycoperdon* spores, one puffball to one cup of hot water and steeped for 15 minutes, are drunk for sore throat, colds, coughs. Anti-tumor and anti-viral activity has been found in *Calvatia*.

## II MORELS

A1 Ridges on mature cap black and pits elongated, brown; cap more or less conic; especially under aspens, though also conifers; on ground in sandy soil,

often in burns a year or two old; in spring when first leaves are out . . . . . . . . . . .
. . . . . . . . . . . . . . . . . . . . . . . . . . . . . Black Morels, *Morchella angusticeps* group.
A2 Ridges light-colored: white, tan or yellowish , , , , ,         , , , , , , , , . . D

B1 Ridges white, pits dark; grassy areas, cultivated ground, wood edges . . . . . . . .
. . . . . . . . . . . . . . . . . . . . . . . . . . . . . . . . . . . . White Morel, *Morchella deliciosa.*
B2 Ridges and pits both pale, tan or yellowish brown, the ridges lighter than the
pits . . . . . . . . . . . . . . . . . . . . . . Common or Yellow Morel, *Morchella esculenta.*

(Plate 1, Fig. 4.) Morels are small to medium sized mushrooms, 1½ ″ to 6 ″ high, with cap and stalk readily distinguishable but grown together where they join, rather than joined at or near the cap apex. Cap and stalk are hollow. The surface of the cap has deep pits separated by a definite network of ridges, not just folds or wrinkles. The flesh is brittle.

Morels grow on the ground especially in spring, in many habitats. In the woods they are hard to see; Black Morels look like fallen pine cones. Species are not well sorted out and there are many intermediates.

All morels are edible but should be cooked. The Black Morel group disagrees with some people, especially in large quantities over several days, or with alcohol. For most people they are a great delicacy. Andrest recommends parboiling twice and discarding the water or else drying before eating for Black Morels. We endorse this recommendation.

**Look-alikes.** The most important false morel is *Gyromitra esculenta,* which has a brown wrinkled irregular cap (not with the honeycomb of pits and ridges of the morels), chambered inside. It is eaten in some places but on the other hand has caused deaths with its poison, monomethylhydrazine (MMH), a substance which is also used as a rocket propellant. Perhaps there is less in some strains or individual mushrooms; perhaps it is safe if cooked exactly right, so the MMH boils off but is not inhaled as vapor; but why fool around? MMH may also be carcinogenic.

Other false morels, many poisonous or of unknown edibility, need not be confused if they are looked at with due attention: they too are wrinkled, folded, irregular, saddle-shaped or smooth, without the clear characteristics of true morels.

**Cooking.** Morels should be cut lengthwise to check for bugs in the hollow interior, but need not be cut finer. They are good sauteed in butter for 5 minutes, with basil and garlic. They freeze and dry well.

**Names.** "Morel", "Sponge-Mushroom", names shared in sound or meaning with other languages: Morchel (German); morille, manigoule, éponge (French); morilla, colmenilla, carraspina, etc. (Spanish); spugnolo (Italian); smorchok (Russian); yangdujun ("Sheep's-stomach Mushroom") (Chinese).

## III TOOTH MUSHROOMS

A1 Cap 2″ to 4″ wide, yellow-orange, tan or near white, margin often wavy, with pale spines of mixed lengths beneath that darken to orange with age or bruising; flesh soft, brittle, pale, mild in taste; on ground or rotten wood in forest, autumn ............... Hedgehog Mushroom, *Dentinum repandum*.
A2 Cap 3″ to 6″ wide, brown, with prominent darker raised pattern like scales; flesh, and spines under cap, grey-brown or tan; spines descending onto top of stalk; center of older caps depressed, sometimes with hole connecting into hollow stalk; stalk short; taste mild to bitter; forest, on ground, late summer and autumn ...................... Owl Mushroom, *Hydnum imbricatum*.

(Plate 2, Figs. 5 and 6.) The two members of the Tooth Mushroom family described here look like normal standard mushrooms, with horizontal cap and vertical stalk, except that beneath the cap they have soft downward-hanging spines, fingers or teeth on which the spores are borne. We find only the vaguest hints of anything unwholesome in the whole family, and according to some authors none are known to be poisonous; but most are woody, leathery, or very bitter.

The Hedgehog Mushroom *(Dentinum repandum)* is variable and widespread, growing all across the northern hemisphere and even in Tasmania. The Owl Mushroom, whose markings look something like the plumage of an owl such as the Long-eared Owl, is sometimes bitter (or may be confused with the bitter *Hydnum scabrosum)*; only mild and young ones are good to eat.

**Look-alikes.** As long as a mushroom has the downward-hanging teeth under its cap it is unlikely to be confused with any dangerous one.

**Cooking.** The Hedgehog Mushroom is a prize edible: sharp clear flavor with slightly buttery or cheesy overtone, mild and delicate (though some strains are said to be bitterish), good texture. To

Fig. 5. *Dentinum repandum* **(Hedgehog Mushroom)**

Fig. 6. *Hydnum imbricatum* **(Owl Mushroom)**

some they are reminiscent of Chanterelles. The quality is best when they are young. They are almost never wormy. A lot of juice is released in cooking. They should be sauteed till not quite crisp, only a minute or two.

Owl Mushrooms are good sauteed fresh, but they are at their best dried and used as a strong rich mushroom flavoring. Snow peas lightly stir-fried with dried Owl Mushroom which has been soaked in water and browned in butter are delicious.

**Names.** Most of the names for *Dentinum repandum* point to the teeth or the golden color: Stoppelpilz or Stoppelschwamm ("stubble-mushroom"), or Semmelpilz (like a browned roll) (German); pied de mouton ("sheep's foot"), barbe de chèvre ("goat's beard"), langue de chat ("cat's tongue") (French); steccherino dorato ("little golden spiny"), gallinaccio spinoso ("spiny Chanterelle") (Italian); yezhevik zholtiy ("Yellow Hedgehog-Mushroom") (Russian); chijun ("tooth mushroom") (Chinese).

The Owl Mushroom is also called Scaly Urchin or Urchin Mushroom ("urchin" is an old word for hedgehog). German says "Hawk Mushroom" Habichtspilz, or "Deer Mushroom" Hirschschwamm; Italian steccherino falso (bruno), "false (or brown) Hedgehog mushroom"; Russian yezhevik pyostryy "motley Hedgehog"; Chinese linxingjun "fish-scale mushroom" or qiaolinrou chijun "projecting scales fleshy tooth mushroom".

## IV BOLETES

A1 Small to medium (cap diameter 2 " to 6 "); soft flesh; on ground under conifers; stalk with more or less straight parallel sides; cap viscid (or with gelatinous layer under felty fibers); may have tiny sticky dots on stalk, especially when old; may have pores arranged in radial rows out from stalk, especially near the stalk, when young (use magnifying glass) *(Suillus)* . . . . . . . . . . . . . . . . . . . . . . . . . . . . B

A2 Medium to large (4 " to 9 " cap diameter); thick solid flesh; on ground under hardwoods and conifers; netted pattern or fibrous dark tufts on stalk; stalk often appearing swollen at middle or bottom; cap dry or slightly slick when wet; tubes randomly arranged; no veil . . . . . . . . . . . . . . . . . . . . . . . . . . . . . . . . . . . . D

B1 Cap 2 " to 4 ", varicolored with layer of red-brown fibers or scales over yellowish viscid surface; yellow pores staining dingy red-brown; yellowish stalk bruising green/dark blue at base, with reddish fibers below white hairy ring (hard to find on old specimens); flesh of cap yellow, staining buff pink; with Douglas fir . . . . . . . . . . . . . . . . . . . . . . . . . . . . . . . . . . . . . . . . . . . . *Suillus lakei* p. 36

*(S. pictus,* under eastern white pine, is similar.)
B2 Cap without fibers, viscid; not staining, no ring . . . . . . . . . . . . . . . . . . . . . . .C

C1 Cap 2 " to 6 ", pale buff to yellowish, sometimes with darker mottling, turning to darker orange cinnamon, viscid; pores off-white to buff, yellowish or light brown in age; cap flesh pale yellow becoming lemon yellow, watery, no stain; stalk white at first without dots, turning brownish, with yellow top and reddish brown dots; no ring . . . . . . . . . . . . . . . . . . . . . . .*Suillus granulatus* p. 36
C2 Cap 2 " to 4 ", dark rich brown or greyish brown when young, paler in age, very viscid, smooth; stalk short, white, yellowing in age, no dots (or occasionally dots on old ones); pores yellow, darkening; flesh white, yellowing; no stain; no ring . . . . . . . . . . . . . . . . . . . . . . . . . . . . . . . . . . . . . . . . . . . . .*Suillus brevipes* p. 37

D1 Stalk swollen, club or egg-shaped and as fat as the cap at first, netted pattern on top one-third; pores small, not bruising or scarcely bruising, white turning yellow to olive in age; cap leathery, smooth, light tan or light brown to yellowish or reddish brown; flesh thick, white, firm, not staining; on ground in pine woods, especially late summer-early fall . . King Bolete, *Boletus edulis* p. 38
D2 Fibrous or scale-like tufts on stalk, dark brown or black at least in age; red-orange or brown-orange smooth or slightly fibrous cap, sometimes slick when wet, often with bits of outer edge of cap skin loose and hanging; flesh white, staining grey-blue or reddish blue, then blackening; pores greyish-white aging to olive and brown; stalk often thicker toward base, tough; young ones shaped like big acorns . . . . . . . . . . . . . . . . . . . . . . . . . . . . . . . . . . . . . . . . . . . . .*Leccinum* p. 40

If there were only the boletes, the whole enterprise would still be worthwhile. These are splendid, flavorful, nourishing mushrooms, widely distributed and often plentiful; they are easy to identify for practical purposes, and none are life-threateningly venomous. They are constructed like the archetypal mushroom--cap, stalk--except that they have no gills under the cap, but vertically oriented tubes instead, whose mouths make an array like a miniature honeycomb or sponge surface. They are forest mushrooms and grow on the ground.

Boletes with red pore mouths should not be eaten as they may cause gastro-intestinal upset, though this may not be true, at least not for everybody, if they are thoroughly cooked. In any case they are less common than good edible kinds.

### *Suillus*

See p. 34, A1. Sounds like "swillus"--here, all agree. Some are called Slippery Jacks.

These medium-sized soft-fleshed boletes are some of the most common; often the woods are full of them. They are good but generally felt to be not as good as King Boletes and *Leccinum,* and many mushroomers leave them if other kinds are abundant. In Russia, though, they are ranked among the half-dozen best edibles. They have a long season and grow in all kinds of conifer forest, so in any case are a good standby.

Any slime should be scraped or washed off the tops. The tubes of older ones should be peeled away, but in younger ones they are good to eat. One trouble with *Suillus* is that they are fussy to prepare; dirt and pine needles stick to the tops, and the flesh is often thin. They soak up water and should be patted dry with a paper towel after washing.

A few *Suillus* may give minor digestive upsets but it would be misleading to call any poisonous, in North America at any rate. Our experience is that *S. tomentosus*--a yellow bolete whose flesh and tubes stain bluegreen--sometimes has a laxative effect. The three given here are among the best for eating. They freeze well, and dry to an attractive weathered dark brown.

### Suillus lakei

See 34, B1 (Plate 2). In season, "look up and see a Douglas fir, look down and see a *Suillus lakei.*" They are often abundant where this tree grows, and beautiful with their yellow stalk and pores, variegated cap, and entrancing stains. They are also firmer-fleshed than other *Suillus*. They have a very long season, from early summer right into late fall.

**Look-alikes.** Hard to mistake for any other; tube mouths yellow, not red, though they stain reddish-brown.

**Cooking.** Slice and saute with basil, a little garlic and a sprinkle of salt till edges brown (2-3 minutes). They dry and freeze reasonably well.

### Suillus granulatus ·

See p. 35, C1 (Plate 3 ). Grows under various conifers. We find it especially under lodgepole pine in August and September.

**Look-alikes.** *S. granulatus* has dots on stalk, no ring, does not stain blue. It could be confused with *S. punctatipes,* which has similar colors but pores with more radial elongation, and is also good to eat.

**Cooking.** Must be fresh and young; doesn't keep well, and worms like it. Mild delicate flavor. Try sauteed with butter, garlic and tarragon.

**Names.** French: bolet granulé, cèpe jaune des pins ("yellow pinebolete"); Italian: boleto granuloso, fong delle vacche (dialect: "cow mushroom"); German: Körnchen-Röhrling ("bolete with little grains"), Rotzling, Pimk, etc.; Russian: maslyonok zernistyi ("slippery or butter-mushroom with grains"); Chinese: dianbing niangai niuganjun ("spotted stalk sticky cap beef liver mushroom"). The "grains" in many names, including the Latin one, are the dots on the stem.

**Miscellany.** A favorite in Europe; in Russia grows in Siberia, the Urals and elsewhere but especially in the Caucasus; also found in Australia, the Philippines and Madagascar.

### Suillus brevipes

See p. 35, C2 (Plate 2). With two and three needle pines in many parts of North America; grows also in China. Mostly in August and September in the Southern Rockies.

**Look-alikes.** With no ring, short white stalk without dots (at least in young specimens), smooth viscid brown cap, it is unlikely to be taken for another species.

**Cooking.** Soon wormy, especially in warm weather; it is important to check and clean immediately on picking. The slimy cap peels easily when moist. Sautes well with butter and dill and fines herbes. Good sauteed either lightly, semi-crisp, or crisp (when it is something like potato chips); alone or with green vegetables such as string beans.

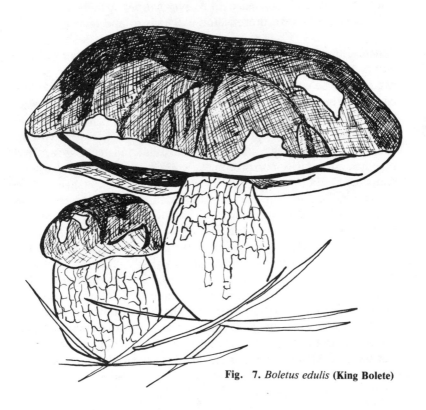

Fig. 7. *Boletus edulis* (King Bolete)

## KING BOLETE, *Boletus edulis*

See p. 35, D1 (Plate 3 , Fig. 7). This is a complex species, widespread and variable in stature and cap color. The flesh does not stain; the cap is not viscid or only slightly so; the stalk is stout and has neither speckles like most *Suillus* nor dark tufty scales like *Leccinum*--usually a light netted pattern can be made out; the tube mouths are not red or orange, and it is a big mushroom, one of the biggest.

King Boletes are not exactly rare but they are idiosyncratic and will grow neither every year nor in every likely place. A few at least can usually be counted on, but in the occasional bonanza year you can pick them till you're tired of stooping, clean them till your fingers hurt, eat them till they're coming out your ears, and still have plenty to freeze and dry. The young ones, appropriately, are shaped like champagne corks. The King Bolete is the most prized mushroom of the northern hemisphere forests where it grows.

**Look-alikes.** There is virtually no danger of confusion with dangerous species provided the mushroom is looked at with due attention: tube mouths are not red, flesh does not stain. The taste is not bitter, but mild and agreeable.

**Cooking.** King Bolete has a subtle taste that does not pall. It is a delicate, yet solid and dependable flavor, described by some as slightly nutty, that one wants to come back to again and again. It is like something remembered from a former life, somehow familiar even on first acquaintance--"ah yes, that!" It carries flavors well, can be cooked all ways, goes with anything. Young ones, as usual, are best. Young tubes should be left on as they hold much flavor; but older tubes need to be peeled. When King Boletes become soft and flabby they are too old to gather.

They take excellently to both drying and freezing. Dried ones can be soaked 15 minutes in hot water then sauteed with butter, garlic and salt; they are different from the fresh, but delicious. Or dried slices without soaking can be put on pizza, under the cheese, and are chewy and tasty. Any general mushroom recipe is at its best with the King Bolete.

**Names.** Russian: belyy grib "white mushroom" because it stays white when dried; or borovik, belovik, zhatnik, glukhar' ("wood-grouse"), korovyak, pechura, medvezhatnik ("bear-hunter"), struyen', tolkach ("push-engine, go-getter, fixer").

Italian names are listed on p. 76 . In French it is cèpe, gros pied ("bigfoot"), grosse queue ("bigtail"), potiron ("pumpkin"), etc.

German has Steinpliz ("stone mushroom"), and many other names: Dobernigel, Doberling, Herrenpilz, Steinkopf, Steinbott, Braunkopp, Gschlachter, Küefotzen....The Chinese name is meiwei niugan ("fine-tasting beef liver'").

**Miscellany.** The myriad names reflect the esteem accorded the King Bolete in Eurasia. A Russian writer (Andrest) calls it the dream of all mushroomers: "They return unsatisfied from the forest unless there are at least a few King Boletes in their basket. Often this determines how successful the excursion into the forest has been. The mushroomer is always met at home with the question: How many 'whites'?"

Traditional Chinese medicine says it relieves colds and joint pains; Japanese research shows activity against influenza virus and tumors.

The King Bolete is coming to be prized and sought in North America too. It is fun to check the prices of imported ones now and then in specialty food stores. It is possible to pay ten dollars or more a pound for the best quality dried "cèpes".

### Leccinum

See p. 35, D2 (Plate 4 , Fig. 8). These are large, fleshy and ubiquitous forest mushrooms with a long season, spring to fall, and are fine edibles and keepables (although generally considered not quite as good as King Boletes). Clearly boletes by the small tubes under the cap, they are distinguished by the blue-grey, reddish-blue, and finally black staining (in most species) of the originally white flesh; and especially by the scabers, or fibrous tufts, that stand out from the stalk. These may be white on very young mushrooms but darken with maturity. The stalk is hard. The pores are dull white, turning olive brown. The caps are greyish to brown, red-brown or cinnamon, or orange.

In the southern Rockies the common ones are *Leccinum aurantiacum* and *L. insigne,* with very little difference except that the latter grows only with aspen. *L. scabrum* has a greyish to brown cap, doesn't stain, and is found with birch.

**Look-alikes.** The species or varieties of *Leccinum* are hard to distinguish among themselves, but all are good edibles. The genus

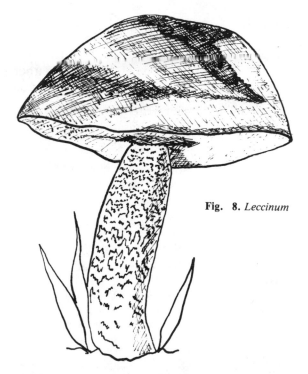

**Fig. 8.** *Leccinum*

is easy: find the scabers on the stalk, and avoid red tube mouths.
**Cooking.** Must be sauteed lightly; will not crisp well, but turns
fibrous if overcooked. Our orange-cap Leccinums turn black when
cooked and dark tan with inky tints when dried. The fresh
mushroom is good sliced thin and sauteed slowly with butter, sage,
marjoram and a sprinkle of salt 7 to 10 minutes or until it just starts
to brown. Thyme is good; oregano is not.

Dried *Leccinum* has a difficult texture, and is better in a mixed
dish than alone. Soaked 15 minutes in hot water, then stir-fried
with spinach, garlic, soy sauce and some of the soaking water,
finishing covered for 3 minutes, for example, it is very good.

Frozen the texture is less a problem: best used in stir-fried dishes,
spaghetti sauces, pizza, or other preparations in which texture con-
trasts are an advantage.

**Names.** French names for various Leccinums include bolet orangé,
cèpe roux ("red [King] Bolete"), bolet rude, raboteux, roussin. An

Italian word for *L. scabrum* is porcinello ("little porcino, i.e., lesser King Bolete"). Russian has podosinovik ("under aspen") for *L. aurantiacum,* and podberyozovik ("under birch") for *L. scabrum.* They are considered one of the best mushrooms in Russia and are some of the most frequently gathered. For *L. aurantiacum* Russia, also has krasnyy ("red"), krasnushka (ditto), krasnogolovik ("red-head"), kazarushka, bachnyuk; for *L. scabrum:* babka, seryy grib ("grey mushroom"), podgreb, among others.

In German *L. aurantiacum* is mostly also red, "rot": Rotkappen, Rotháuptchen, Rotdocke; also Frauenschwamm ("women's mushroom"). *L. scabrum* is Birken-Röhrling ("birch-bolete"), Geisspilz or Geissfuss ("goat mushroom, goatfoot"), Pfaffenkopf ("priest's-head"), Kapuzinerpilz ("Capuchin mushroom"), and others. *L. aurantiacum* in Chinese is chenghuang youbing niuganjun ("orange scaber-stalked beef liver mushroom").

## V AGARICS

A1 "Gills" low blunt ridges resembling folds, running down stalk a little way, with forks and cross-veins; yellow, gold to orange mushroom; 1 " to 3 " high, cap 1 " to 3 " wide, often with wavy margin; cap and stalk grown together without sharp angle between, vase shape; fleshy; fruity fragrance, mild or slightly peppery taste; on ground in forests, often in clumps . . . . . . . . . . . . . . . . . . . . . . . . . .
. . . . . . . . . . . . . . . . . . . . . . . . . . . . . . . .Chanterelle, *Cantharellus cibarius,* p. 45
A2 Gills sharp-edged, relatively deep, may fork but no cross-veining . . . . . . . . .B

B1Containing a carroty red-orange juice that turns green on exposure to air; cap tawny orange, 2 " to 6 " wide, stalk and gills orange, all developing greenish tints with age; cap with faint concentric pattern, margin inrolled at first, developing sunken center, slightly viscid when young; gills attached or running slightly down stalk; stalk hollow; spore print off-white (cream or buff); flesh tender, crumbly; on ground under conifers . . . . . . . . . . . . . . .*Lactarius deliciosus,* p. 46
B2 No such juice . . . . . . . . . . . . . . . . . . . . . . . . . . . . . . . . . . . . . . . . . .C

C1 Spore print pinkish-brown or salmon; cap brown with grey, red or violet tinge, 1½ " to 3 " wide, flesh soft and white; gills free, light tan or grey becoming pink; radishy or turnipy odor; on rotting hardwood . . . . . . . . . . . . . . . . . . . . . .
. . . . . . . . . . . . . . . . . . . . . . . . . . . .Deer Mushroom, *Pluteus cervinus,* p. 48
C2 Spore print another color . . . . . . . . . . . . . . . . . . . . . . . . . . . . . . . . . . . .D

D1 Spore print dark (brown, black, dark grey) . . . . . . . . . . . . . . . . . . . . . . . .E
D2 Spore print light (white, cream, yellow, pale lilac) . . . . . . . . . . . . . . . . . . . .H

E1 Spore print cocoa-brown (not rusty or reddish brown, cinnamon-brown, or yellow-brown); cap white to light brown or tan, 1 ½ " to 6 " wide, smooth or with fibrous strands, not viscid; gills free, white or pink at first but brown when mature; stalk separating easily from cap, with ring, smooth at base without cup or roughened bulb or ridges of tissue (dig up bottom of stalk in gathering); odor mild, mushroomy or faintly anise-like or licoricy, pleasant; on ground in grass or in forest *(Agaricus)* . . . . . . . . . . . . . . . . . . . . . . . . . . . . . . . . . . . . . . . . . . . . . . . . . . . . . . . G
E2 Spore print black or dark grey. . . . . . . . . . . . . . . . . . . . . . . . . . . . . . . . . . . . . . . . . . F

F1 Cap white with reddish-brown shingly scales, turns to black liquid from margin inwards as it matures; to 6 " high, cylindrical; gills free, hidden under cap, white at first then blackening; stalk hollow, with ring; in clumps on hard ground, in grass, on roadsides . . . . . . . . Shaggy Mane, *Coprinus comatus,* p. 49
F2 Cap purple with darker spots in age, very viscid, 1 " to 4 " wide, flesh white or greyish; gills grey, descending stalk; slimy ring; base of stalk yellow; on ground under conifers . . . . . . . . . . . . . . . . . . Yellow Foot, *Gomphidius glutinosus,* p. 51

G1 In grass; cap 1 " to 4 " wide, white, with or without brown fibers, not bruising or bruising faint pink; gills free, pink at first, turning brown; stalk short, thinner at base, no volva, thin ring; chocolate-brown spores; odor mild, pleasant . . . . . .
. . . . . . . . . . . . . . . . . . . . . . . . . . . Meadow Mushroom *Agaricus campestris,* p. 52
G2 In forest (hardwood or conifer); cap 3 " to 6 ", smooth, shiny, silky, white yellowing in middle, fast light yellow stain when handled, may stain rusty brown later; evanescent anise or licorice odor when handled; gills off-white to pinkish at first, turning brown, free; stalk long, white, sometimes bent, may be flattened or somewhat bulbous at base but no volva or tissue remains, base does not stain yellow; thick ring; brown spores; a big lumpy mushroom . . . . . . . . . . . . . . . . . .
. . . . . . . . . . . . . . . . . . . . . . . . . . . Wood Mushroom, *Agaricus sylvicola,* p. 53

H1 On wood . . . . . . . . . . . . . . . . . . . . . . . . . . . . . . . . . . . . . . . . . . . . . . . . . . . . . . . . . I
H2 On ground . . . . . . . . . . . . . . . . . . . . . . . . . . . . . . . . . . . . . . . . . . . . . . . . . . . . . . . K

I1 No stalk, or a short off-center hairy stalk; cap 2 " to 8 " wide, cream, tan or slate-grey, smooth, sometimes with crinkled edges, fan-shaped; flesh white, thick, soft (though may be tough near point of attachment); gills thick, white or cream, running onto stalk if any; spores white to pale lilac; no ring; on dead hardwood usually in shelving groups . . . . . . . . . . . . . . . . . . . . . . . . . . . . . . . . . . . . . .
. . . . . . . . . . . . . . . . . . . . . . . . . . . Oyster Mushroom, *Pleurotus ostreatus,* p. 54
I2 With a definite central stalk . . . . . . . . . . . . . . . . . . . . . . . . . . . . . . . . . . . . . . . . . J

J1 Stalk thin, 1/8 " to 1/4 ", brittle-fibrous, darkened with reddish-brown to black velvety covering on mature mushrooms; no ring; caps delicate, to 2 " wide, orange-brown, darker in middle, viscid; flesh soft, thin; gills white, attached; spores white; in clusters on dead hardwood. . . . . . . . . . . . . . . . . . . . . . . . . . . . . .
. . . . . . . . . . . . . . . . . . . . . . . . . Winter Mushroom, *Flammulina velutipes,* p. 56

J2 Stalk thicker, ½ " to ¾ ", tough, with cottony ring, white above ring and sometimes reddish-brown below, and yellow tones sometimes on ring and below; cap 2 " to 4 " wide, tawny-brown or honey-brown sometimes with pink tint, slightly viscid, with small upstanding brown hairs or scales especially in middle; flesh white, thick; gills white to brownish, attached; spores white; in clusters on hardwood or conifer, usually dead wood, often very numerous
............................Honey Mushroom, *Armillariella mellea,* p. 57

K1 With a thick soft white ring; cap 3 " to 8 " wide, white and with inrolled cottony margin at first, becoming cinnamon-brown and streaked with fibers, hairs or scales, especially in middle; flesh thick, white; attached close gills, white, bruising red-brown; thick short hard stalk, sheathed below with soft white veil tissue turning to reddish-brown scales; spores white; distinctive odor, not like mushrooms in general but fragrant, exotic, spicy, willowy; under lodgepole and other 2-needle pines ............Pine Mushroom, *Armillaria ponderosa,* p. 59
K2 Without ring .....................................................L

L1 Cap 2 " to 6 " wide, red, yellow, green, orange, brown or white, smooth, skin peeling easily; flesh dry and crumbly, does not exude juice when cut; gills white or near-white, at least when young, attached to nearly free, delicate, crumbling easily when rubbed; stem breaks like soft chalk; no ring; spores white or yellow; on ground in woods ......................................*Russula,* p. 60
L2 Cap 1½ " to 4 " wide, yellow to greenish-yellow with tan to ruddy-brown middle, smooth, viscid; flesh firm, white to yellowish, does not exude juice when cut; gills yellow to light greenish-yellow, adnexed, close; stalk light yellow, short; no ring; spores white; odor mild, mealy; half-buried in sandy soil under pines ...................Man-on-Horseback, *Tricholoma flavovirens,* p. 62

The agarics are fleshy mushrooms with cap and gills and (usually) a central stalk, the archetypal Mushroom like the familiar *Agaricus bisporus* of American markets. Though they are the most numerous and most complicated group, there is nothing mysterious about identifying those included in this book; and though a few agarics are seriously poisonous, there is no reason to err if the descriptions are checked closely. Those given here are a small fraction of the hundreds of species a gatherer will encounter, but they are among the best, the safest, the commonest, and consequently, the most often consumed in North America and around the whole northern forest zone. The ones excluded are by no means all poisonous: some are good, most are harmless. But many are scarce, small, incompletely researched, or hard to identify without microscope and chemicals.

**Fig. 9.** *Cantharellus cibarius* **(Chanterelle)**

## CHANTERELLE *Cantharellus cibarius*

See p. 42, Al. (Plate 4, Fig. 9). The best places to look for Chanterelles in the southern Rockies are above 9000 feet in stands of lodgepole pine, aspen, and Douglas fir, where kinnikinnik and blueberry grow, not on south-facing slopes, in autumn. The color is like fallen aspen leaves, apricot, peach, egg-yolk, with gills slightly lighter than cap and peeling easily (though they needn't be peeled before eating). Chanterelles have a pleasant fruity scent somewhere between ripe apricot and wet rag. The meaty cap is flat or slightly depressed in the center, and may have little dents. They are one mushroom that will not fall apart even if collected in a paper bag.

**Look-alikes.** The Jack-O-Lantern Mushroom, *Omphalotus olearius (Clitocybe illudens),* is orange and vase-shaped but has sharp-edged, narrow, close gills without forks or cross-veins; it grows on wood, but the wood may be buried so that it seems to be growing on the ground. It produces gastro-intestinal disturbance but is not deadly.

*Hygrophoropsis aurantiacus (Clitocybe aurantiaca),* the False Chanterelle, is an exceedingly beautiful orange mushroom that grows on the ground or on rotten wood, and is similar but its gills are close, narrow, and have numerous forks but not the cross-veining of the Chanterelle. Reports conflict, but it seems to be edible at least for some persons, though not very good. Upsets blamed on it may have been caused by the Jack-O-Lantern.

There is virtually no danger of mistaking a really deadly mushroom for the Chanterelle. Neither of the look-alikes have its fruit aroma. The key is to look closely at the gills.

**Cooking.** The aroma and color are pleasant to work with, and Chanterelles are less watery than most mushrooms and lose less in cooking. The taste is delicately fruity-buttery, the texture smooth. Good sliced and sauteed one or two minutes, till not quite crisp.

The prime delicacy is Cream of Chanterelle Soup (see p. 100). The flavor and texture is especially fine and light, and it should be served at the beginning of a meal.

Chanterelles can be dried but they are better frozen, remaining almost as good as freshly picked ones.

**Names.** Chanterelles are well known and much gathered in the temperate zones of Eurasia. The main German name is Pfifferling, referring to a peppery taste some Chanterelles have; also Eierschwamm ("egg-mushroom"), Rehgeiss ("roe deer"), Nagerl ("small nail"), Schweinfüsserl ("little pig's-foot"). The French word is chanterelle, also girole, oreille de lièvre ("hare's ear"), crête de coq ("coxcomb"), jaunette ("little yellow"). Spanish is rebozuelo or rusiñol; Italian gallinaccio, capo gallo and many dialect names such as gallettu, finfer, fonz zald, cresta de gallo. In Russian it is lisichka ("little fox") or petushok ("cockerel"). A Chinese name is jiyoujun ("chicken-fat mushroom").

*Lactarius deliciosus*

See p. 42 , B1 (Plate 5 , Fig. 10). Now regarded as a group of species and varieites, some better eating than others but all with the orange to green color change and perfectly wholesome. Mid-July to mid-September in the southern Rockies. One of the marvels of the woodland, with its extraordinary colors. But insect larvae soon find it and it must be harvested as soon as possible after it comes up. Often plentiful.

**Look-alikes.** *Lactarius* the genus is named for the "milky" (watery in some) juice. A slant cut across the gills should show droplets, though the juice may be hard to find in older or dry specimens. This characteristic, with attached gills, no ring, white to buff spore print, no volva, and location on ground in forest, should preclude confusion with other genera.

There are many common species of *Lactarius* (though none so frequent here as *L. deliciosus)* and they are a real sight with their varied latexes, which may or may not turn color on exposure to air. Most authorities recommend avoiding acrid (stingy) or bitter species, and "all those in which the latex is white at first but changes to or stains the flesh violet, reddish, lilac, or some shade of yellow" (A.H. Smith, in Rumack).

*L. deliciosus,* with its peculiar orange to green color change, is unlikely to be confused with any other mushroom.

**Cooking.** The mushroom varies in its delectability but many English-language writers register their disagreement with its name *deliciosus,* calling it overrated, or only fair. To some tastes,

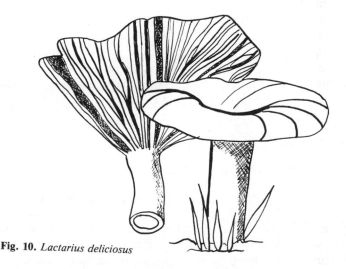

**Fig. 10.** *Lactarius deliciosus*

though, there is no finer mushroom. The texture is dry and crumbly, but tender; the flavor is superb. Slice and saute 5 minutes with butter and a little salt; coriander and cumin go well.

To freeze, first saute 1 minute or until juice is released. Defrosted in a hot skillet with a little butter and salt, it is excellent.

*Lactarius deliciosus* loses flavor and texture when dried and is not good in most recipes. It does make an excellent chutney: rehydrate and lightly brown, together with raw sunflower seed, in butter; mix with chopped pimento-stuffed green olive.

**Names.** Italian: lapacendro buono; it is depicted in the frescoes of Pompei. Spanish: níscalo or mízcalo, robellón. French: barigoule, orangé, roussillon, vache rouge ("red cow"). German: Edelreizker or echter Reizker ("noble, or genuine, Reizker"), Karottenmilchling ("carroty milk-mushroom"). Russian: ryzhik; gruzd' is *Lactarius resimus,* and the name is also applied to some other *Lactarius* (as well as some *Russula*). Chinese: sung rugu ("pine milk-mushroom").

**Miscellany.** Grows across North America, Europe and Russia, and is highly valued. "Skillfully salted *Lactarius deliciosus*," says Andrest, "retains the resinous aroma of pine-needles and the freshness of the forest....Formerly inhabitants of the Urals [the low mountains west of Siberia], when they found a large fruiting...salted them down right in the woods. For this they brought...cedar kegs which had been steamed with juniper berries, and carefully packed the mushrooms in tight rows, sprinkling them with coarse salt, first having wiped each mushroom with a linen cloth. They added no spices. The famous salted *Lactarius deliciosus* of the Urals was distinguished by its inimitable scent of cedar, pine pitch and spruces moist with dew."

## DEER MUSHROOM *Pluteus cervinus*

See p. 42, C1 (Plate 5). Another collective or variable species, widespread, and a good (though not superior) eating mushroom, quite easy to recognize. The cap is lighter around the edges than in the middle, and may have darker fibrous scales. The gills are close, tan or grey turning pink; spores are pinkish. The stalk is light tan to pinkish-brown and has no ring or volva.

**Look-alikes.** The genus *Entoloma* can cause gastro-intestinal disturbances, and also has pinkish spores; some may appear on rotten wood. But gills are attached in *Entoloma,* though in older specimens they may pull away from the stalk.

The deadly *Galerina autumnalis* grows on wood, but it has a ring, yellowish attached gills, and a rusty brown spore print.

Other species of *Pluteus* that resemble the Deer Mushroom enough that they might be confused with it are just as edible: *P. atromarginatus* has nearly black cap and black gill-edges; *P. magnus* has darker, wrinkled cap.

**Cooking.** The Deer Mushroom does not seem to be the main goal of many mushroom excursions; but young ones, sauteed in butter, have their own good distinctive flavor, and it is quite common.

**Names.** French: plutée; Italian: pluteo cervino; Russian: plyutey oleniy ("deer mushroom"); Chinese: hui guangbinggu ("grey bare-stemmed mushroom").

**Miscellany.** The Deer Mushroom is found in Australia as well as North America and Europe. It is in the same family as the Chinese caogu ("straw mushroom", *Volvariella volvacea*), cultivated in the warmer parts of East Asia, India and Africa.

## SHAGGY MANE *Coprinus comatus*

See p. 43, F1 (Plate 5, Fig. 11). An available mushroom, one that you don't have to go to the woods for. It may appear in spring and summer but the biggest fruitings in the southern Rockies are in late September and October, even after snow. It is a very distinctive black-spored mushroom, growing in big bunches that can lift clods of hard earth. Rarely it grows in grass in the woods, but it is usually in the open. The autodigestion, whereby it turns itself to black ink, is weird and people shy away from this mushroom at first.

Shaggy Manes must be gathered at once. Only pure white ones are good, and once they start turning black the flavor is spoiled and they can cause digestive upsets. When they appear it's time to drop everything and pick quantities to eat and freeze.

**Look-alikes.** Black spores, the turning to black liquid, large size, cylindrical or long egg-shape, reddish-brown shingly scales on cap,

Fig. 11. *Coprinus comatus* (Shaggy Mane)

make the Shaggy Mane easy to identify. Other vaguely similar poisonous or hallucinogenic mushrooms of the same family grow on manure. *Coprinus atramentarius* is smaller, usually grows on or near dead wood, and has a radially lined grey-brown cap; it also autodigests, and is edible and good but some persons react badly upon drinking alcohol within several hours or even a couple of days after eating it, or if they have had much alcohol shortly before.

**Cooking.** Fresh Shaggy Manes do not keep well more than a few hours even refrigerated; if picked very young and fresh in the morning they can be eaten for dinner that night. They must be kept upright in a basket or pan and handled carefully till cleaned. The caps easily split and the stems come loose, and once the grit gets into the gills it is impossible to remove it all. Hold them upright under a faucet to rinse, and rub the dirt off before trimming the base of the stalk. Water too should be kept out of the gills or they become soggy.

Shaggy Manes are generally agreed to be particularly outstanding with omelet or scrambled egg. They are strong-flavored and need to be well cooked. They are also recommended for sauces and with fish and meats.

Frozen *Coprinus* is only so so. Blanch whole 3 minutes, quick-freeze on trays, then package. The blanching deactivates the blackening enzyme.

**Names.** Inky Cap and Lawyer's Wig are other English names. *"Coprinus"* refers to dung, and many other names do as well, either because of where it or its relatives grow, or because of what it turns into. Russian: navoznik belyy ("white dung-mushroom"--navoznik also means "dung-beetle"), navoznik lokhmatyy ("shaggy dung-mushroom"). German: Schopf-Tintling, Porzellan-Tintling ("tufted, or porcelain, ink-mushroom"); Italian: coprino chiomato ("shaggy Coprinus"): French: coprin chevelu (ditto), goutte d'encre ("ink-drop"), fusée ("rocket"), escumelle. Chinese: maotou guisan ("hairy-headed devil's-umbrella").

**Miscellany.** A 16th century Chinese work by Pan Zhiheng says that Shaggy Mane, or some other *Coprinus,* was called Dog Piss by children. Such mushrooms are good, he says, to treat itchy sores: dry, powder, mix with fat, and apply as an ointment.

## YELLOW FOOT *Gomphidius glutinosus*

See p. 43, F2 (Plate 5 ). The complementarity of purplish cap and yellow foot make this a handsome mushroom, and with its black spores and gills that run down on the stalk it is easy to recognize. It prefers rich ground with plenty of humus, and is common in the southern Rockies in moist Augusts and Septembers.

**Look-alikes.** Might be confused with other *Gomphidius,* but all are edible. *Chroogomphus* has yellow, pink or orange cap flesh, rather than white. It is also edible.

**Cooking.** A pleasant mushroom sauteed with butter and seasoned with oregano, though not to everyone's taste. The texture is soft, and sometimes a faintly medicinal taste is noticeable. The slimy cap skin peels easily and should be removed.

**Names.** Most refer to the yellow foot or the slimy cap. Italian: chiodetto ("little nail") and dialect gambe zalde ("yellow legs"); French pied de rhubarbe ("rhubard foot"); German grosser

Gelbfuss ("big yellowfoot"), Kuhmaul ("cow's mouth"), Schafsnase ("sheep's nose"); Russian mokrukha ("wet" mushroom), sliznyak ("slug").

## *AGARICUS*

See p. 43, E1. More than any others these are the quintessential edible mushroom at least to Americans because they include *Agaricus bisporus,* the mushroom of the markets. Look closely at one of these to see the basic characters of the genus: gills free from stalk, light at first and turning dark with maturity as the brown spores color them; a veil stretching between cap edge and stalk that breaks and adheres to the stalk as a ring; stalk separating cleanly and easily from the cap. The base of the stalk is cut away on the commercial mushroom but should always be checked in wild ones to be sure there is no volva remnant that would mark it as a possibly poisonous *Amanita.*

Many *Agaricus* are hard to tell apart, being variable species or species complexes. A number of the one or two hundred North American species cause gastro-intestinal disturbances in some people (including, rarely, the commercial mushroom). The two given here are edible for most people but as always it is wise to go slow, especially with the Wood Mushroom.

### MEADOW MUSHROOM *Agaricus campestris*

See p. 43, E1 and G1 (Plate 6 , Fig. 12). This is probably the one wild mushroom that is gathered most in America. It is common in many kinds of grassy area, natural or manmade, and is delicious. **Look-alikes.** The principal danger is confusion with poisonous Amanitas which have WHITE gills and spores, and a VOLVA or its remnants at the base of the stalk. Meadow Mushrooms have PINK to BROWN gills and BROWN spores, and no volva. Take care with unexpanded buttons.

Some other *Agaricus* with unpleasant chemical or metallic odors and tastes, perceptible raw or when cooking, are likely to cause gastro-intestinal disturbances. Avoid *Agaricus* with yellow-staining stalk flesh.

*Chlorophyllum molybdites* is somewhat similar and causes gastro-intestinal disturbance to many or most people, but it is larger, and has white to grey-green gills, grey-green spore print, and scaly top. *Leucoagaricus naucinus (Lepiota naucina)* causes upset in some persons, but its gills are white at first, and its spore print is white.

**Fig. 12.** *Agaricus campestris* **(Meadow Mushroom)**

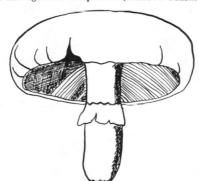

**Cooking.** A fine mushroom, best while the gills are still pink; can be used in the same manner as its store-bought relative.

**Names.** Also called Pink-bottom, to remind gatherers to check the gills. French: agaric des prés ("meadow agaric"), champignon de rosée ("dew mushroom"), boule de neige ("snowball"), rosé des prés ("meadow pink"); Spanish: hongo campesino ("meadow mushroom"); Russian: shampinyon obyknovennyy ("common agaric"), pecheritsa. German: Wiesen-Egerling, Weidling, Brachschwamm, all referring to pasture or ploughland. Chinese: mogu ("mushroom").

**Miscellany.** Anti-viral activity has been found in the Meadow Mushroom.

## WOOD MUSHROOM *Agaricus sylvicola*

See p. 43, E1 and G2 (Plate 6 ). This again is regarded as a group of mushrooms, a variable species or a species complex. It is typically found on level ground in mature forest with spaced trees and rich soil. Most persons can eat the Wood Mushroom without ill effects but for some it, or one or another of its variants, causes gastro-intestinal disturbance. It is delicious, common--in our area it is the most often found forest *Agaricus*--and large, and it keeps well (worms avoid it), so it is worth trying, starting as always with a small bite or two to see how it sits.

**Look-alikes.** Same as for the Meadow Mushroom: watch out for *Amanita,* which also grows in the woods. If the gills are white, if there is any sign of extra tissue around the foot of the stalk, don't eat it. Spores must be brown not white. Also, do not eat "Wood Mushrooms" that have or develop any unpleasant chemical odor or taste, or whose stalk base turns bright yellow when cut.

**Cooking.** Good sauteed, usable in many ways.

**Names.** Grows in Europe, China and Australia as well as North America. French: agaric anisé, rosé des bois. Chinese: pai lin digu ("white forest ground-mushroom").

## OYSTER MUSHROOM *Pleurotus ostreatus*

See p. 43, I1 (Plate 6 , Fig. 13). Also a species complex or a variable species. Grows on many hardwoods, in our area especially on aspen and cottonwood; a real delicacy, and sometimes abundant; does not necessarily require a trip to the woods, in fact may be of better quality in more open places. Has a long season, spring through fall. Must be gathered soon as it is rapidly infested by cute little beetles with a red pronotum, and their larvae.

**Look-alikes.** Unlikely to be mistaken, with stalk missing (or short and off-center), growth on hardwood, thick gills, white spores, smooth cap, thick soft flesh.

**Cooking.** Taste and texture are something like an oyster: "The camel," wrote McIlvaine at the turn of the century, "is gratefully called the ship of the desert; the oyster mushroom is the shellfish of the forest." Superb just sauteed; freezes and dries well.

**Names.** The many names testify to appreciation all around the northern hemisphere. Italian: agarico ostreato, curena, and at least 18 dialect names. French: pleurote en huître ("oyster *Pleurotus*"), oreille de noyer ("wanut ear"), oreille de chat ("cat's-ear"), poule de bois ("wood chicken"). Russian: veshenka obyknovennaya or ustrichnaya; Chinese; hangu ("oyster mushroom"), ce'er ("side ear").

**Miscellany.** Cultivated commercially in Europe and Asia.

**Fig. 13.** *Pleurotus ostreatus* **(Oyster Mushroom)**

## WINTER MUSHROOM *Flammulina velutipes*

See p. 43, J1 (Plate 6, Fig. 14). Despite the name, a summer fruiter here in the southern Rockies; but experiments have shown that it withstands freezing and can then start growth again in a warm spell, when most mushrooms have disappeared. In the

**Fig. 14.** *Flammulina velutipes* **(Winter Mushroom)**

eastern United States and in Europe it is associated especially with elm, in our area with aspen and cottonwood.

**Look-alikes.** The velvety stalk is very distinctive. It has no ring. The very poisonous *Galerina autumnalis,* which also grows on wood, has a ring, and its spores are brown (not white).

**Cooking.** Sauteed it has a good texture and a special clear bright taste. The stalk is too tough to eat. Winter Mushroom has a small amount of cardiotoxic protein which disappears on cooking, so it particularly should not be eaten raw.

**Names.** Also Velvet Foot in English; German has these same two names: Winterpilz and Samtfuss-Rübling. Italian: agarico vellutato, fungo dell'olmo ("elm mushroom"). French: collybie à pied velouté *("Collybia,"* another genus name for it, "with velvet foot"); Russian zimniy grib or zimniy opyonok ("winter mushroom, winter honey-mushroom"). Chinese donggu ("winter mushroom"); Japanese enokitake ("hackberry-tree mushroom"). **Miscellany.** Found in Eurasia, North America, South America, Africa and Australia; cultivated in Taiwan and especially Japan, on sawdust mixed with rice bran. Cultivated ones look different from wild, and are not as good.

## HONEY MUSHROOM *Armillariella mellea*

See p. 44, J2 (Plate 7 , Fig. 15). Still another variable species or species complex. May be variously colored, from honey-brown through other browns and even pinkish (not orange). The upstanding small hairs or scales on the cap may be hard to find, the cap itself may be sticky or dry, the stalk thick or thin, short or long, and the ring may fall off. It is only common sense to gather just those one is sure of, and leave the doubtful ones. The mushroom often fruits by the bushel and there is no need to go without. In our area, it flourishes from mid-August to late September.

**Look-alikes.** The main important chance of confusion is with *Galerina autumnalis,* but this very poisonous mushroom has brown spores, a thin membranous rather than cottony ring, and tawny or brownish gills: the Honey Mushroom's gills are quite white when young, though they stain brownish later. *Galerina autumnalis* is smaller, has a smooth cap, and lacks the upright scales of the Honey Mushroom.

*Pholiota* grows in clusters on wood, but has brown spores. *Gymnopilus* also clusters on wood, but its spores are rusty orange. Neither genus is deadly but both have mild poisons in some species. *Naematoloma fasciculare,* which also clusters on wood, is smaller with smooth cap, greenish-yellow gills, and purple-brown spores.

Spore color is critical. Since the Honey Mushroom grows in groups one can always find a cap tucked under the gills of another so as to register a white print (see color illustration). A definite cottony ring is also important; never mind specimens in which it may have worn off, they're too old to be good eating anyhow. Be leery

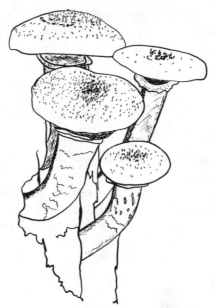

Fig. 15. *Armillariella mellea* (Honey Mushroom)

of guessing that mushrooms on the ground are really attached to buried wood.

**Cooking.** Although the Honey Mushroom is regularly called a choice edible its range of application is actually rather narrow. It should not be eaten raw or undercooked as it often is bitter and burning and disagrees with people. Only young unweathered mushrooms should be eaten. Saute them slowly, in butter, uncovered; they turn sweet and good. (Honey mushrooms are said to make a good baked mushroom loaf.)

Dried Honey Mushrooms are nothing--an unpleasant metallic taste. They can be frozen by first sauteeing in peanut oil with soy sauce added while cooking, then quick-freezing and bagging. These are suitable for Chinese stir-frying, used like dried and rehydrated Chinese mushrooms.

Best of all to our taste, and ideally suited to the immense harvests that frequently offer themselves, is pickling; pickled young Honey Mushrooms are a delicacy second to none (p. 109 ).

**Names.** Italian chiodini ("little nails" which the buttons do resemble), famigliola or famigliola buona ("little family, Holy Family")--one of best known and most gathered in Italy. French: pivoulade, souchette (souche = stump), tête de Méduse ("Medusa's head"); German: Hallimasch (see p. 75 for other names); Russian: opyonok osseniy ("autumn opyonok"), regarded as one of the best edibles; Chinese: mihuanjun ("honey ring mushroom").

**Miscellany.** The copious fruitings are supported by active and aggressive feeding on trees, dead and even living, by means of rootlike strands of mycelium that sometimes glow in the dark. These strands have been found growing over 800 feet deep in an abandoned copper mine in Michigan in tangled masses the miners call "fishnet."

## PINE MUSHROOM *Armillaria ponderosa*

See p. 44, K1 (Plate 7, Fig. 16). A large, prize mushroom quite unlike others in flavor and texture. In our area it grows under lodgepole pine above 9,000 feet from August into October. It is found also in the northeast, north central, and northwest United

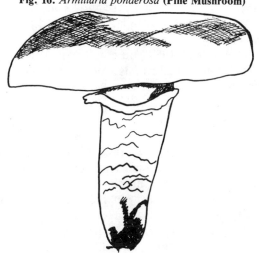

**Fig. 16.** *Armillaria ponderosa* **(Pine Mushroom)**

States. The characteristic willlowy-spicy fragrance is strongest at the gills.

**Look-alikes.** Unlikely to be confused with anything bad: has ring, attached gills, no volva. *Armillaria caligata* has a scent but is not white when young; it is edible. *A. zelleri* has a viscid, orange-brown cap with white margin, and an odor of corn silk. It is not good but not poisonous. *Catathelasma* is an edible genus. The caps are white to light grey or brown, the stalk is narrowed toward the base,and the ring is double. None of these have the distinctive scent of *A. ponderosa*.

**Cooking.** The aromatic (yeasty, perhaps?) flavor and firm texture are unique. This is one mushroom that does not do well simply sauteed in butter. Oriental styles are best: for example, slice 1/8" thick, soak in half soy sauce and half water, broil over charcoal or wood or in oven.

Pine Mushrooms freeze and dry well.

**Names.** The Japanese matsutake ("pine mushroom") is *Armillaria matsutake,* considered by some the same species as *A. ponderosa* (and *A. caligata).* Japanese in America call *A. ponderosa* "white matsutake." The Chinese songxun ("pine mushroom") grows under red pine, a two-needled pine of N. China, Korea and Japan. *Armillaria matsutake* is called songkoumo in China. *"Ponderosa"* means "weighty" and does not refer to the ponderosa pine. *"Armillaria"* comes from Latin armilla, "bracelet", referring to the ring. *A. caligata* is also found in Great Britain.

**Miscellany.** In the fall of 1981 you could pay as much as $7.95 a pound for fresh Pine Mushrooms, gathered from the wild, at Pacific Mercantile in Denver.

*Russula*

See p. 44, L1 (Plate 8, Fig. 17). In this case only one pronunciation is heard: RUSHuhluh. Russulas are common and easy to spot on the whole, with their very crumbly flesh and gills, no ring, and white, creamy or yellow spores; but many individual species are hard to identify without microscopic study and even then the classification is not all sorted out. As a practical matter, many people in many lands eat a number of kinds without ill effect.

Numerous cautions and rules have been set up about not eating red ones, green ones, bitter ones, acrid-stingy ones. According to A.H. Smith, who has studied the subject for decades, the two recommendations that still stand are: do not eat *Russula* raw, and do not eat white Russulas that stain black. Even the portentously named *R. emetica* may be tolerated by some people if thoroughly cooked, or, better, parboiled and the water discarded, then re-cooked. We cannot vouch for it. But in any event, no *Russula* is dangerously poisonous.

Our local preference, pictured here, is *R. decolorans,* common in in the southern Rockies and easy to recognize: reddish-yellow cap, somewhat viscid; creamy yellow-buff spore print; flesh and gills turning slowly grey on bruising or with age, the cut stem greying faster.

Russulas are long-seasoned and one of the most abundant groups of forest mushrooms, but not a general favorite. Only occasionally does one find signs of a true *Russula*-lover's passage: the careful disks of stem, cut to find how high the worms have progressed, and whether the cap is salvageable....Like *Suillus,* Russulas are merely good back-up mushrooms to most gatherers.

**Look-alikes.** As always, it is important to be sure there are no volva remnants and no ring, to preclude mix-up with *Amanita. Lactarius* is similar but has milky (or watery) juice; however, this often dries up.

Many other white-spored genera have a superficial resemblance to *Russula* in that they have no ring and grow on the ground in the forest: *Tricholoma, Clitocybe, Collybia, Leucopaxillus* and *Hygrophorus.* Some are edible and some are poisonous. What is most distinctive in *Russula* is the dry, crumbly, brittle texture which can be fairly easily recognized after some practice in bending, breaking, rubbing, and throwing mushrooms. The fruiting body is rigid, and the gills have an even appearance.

*R. decolorans* is not hard to identify and we recommend sticking to this species until experience is acquired.

**Cooking.** *Russula decolorans* looks beautiful sauteed, with hand-some gold-brown and reddish tones. It remains crisp but is not tough or fibrous and an unkind diner might call the texture

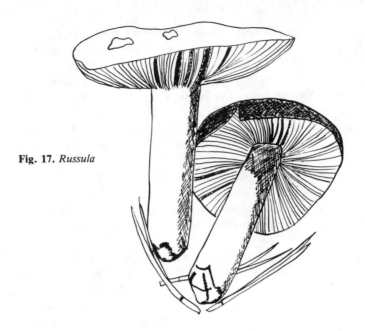

**Fig. 17.** *Russula*

plasticky. The cap is better than the stem.

**Names.** German: Täubling; French: russule; Russian: syroyezhka ("edible raw": not a good idea) or in the Urals: sinyavki. *R. decolorans,* in Chinese: tunse honggu ("fading red mushroom").

**Miscellany.** Inhabitants of the Urals are said to be real Russulophiles.

Says Andrest, evidently not himself a Urals native, "Russulas are the friends of mushroomers, especially of those who cannot seek out the *Lactarius* hiding among leaves and pine needles, who have not the patience to circle the birches in hopes of finding a King Bolete"--which about sums it up.


## MAN-ON-HORSEBACK *Tricholoma flavovirens*

See p. 44, L2 (Plate 8 ). In the southern Rockies fruits in August and September, especially under lodgepole but also with Douglas fir and spruce. Abundant, but only the youngest are free of worms.

**Look-alikes.** No ring, no volva; yellowish gills and stalk distinguish it from most other *Tricholoma,* but if it smells like coal-tar and has a dry cap, it is *T. sulphureum,* unappetizing and perhaps poisonous.

**Cooking.** Some trouble to clean, as it is low to the ground and involved with the sandy soil it favors. Excellent sauteed—the texture holds up well, nice-looking, very good flavor. Marjoram goes well with it. Dries well. To freeze, first saute one minute; excellent put in pan, not thawed, and re-sauteed till crisp.

**Names.** Another Latin name is *Tricholoma equestre, "equestre,* belonging to a horseman or knight, from distinguished appearance," says McIlvaine. German: Grünling (from grün, "green") or echter Ritterling ("genuine knight-mushroom"); French: chevalier ("knight"), canari ("canary"), pied d'âne ("donkey's foot"); Russian: zelenushka ("the green one"). Chinese: youkoumo ("oily mouth-mushroom").

### Amanita verna

I dreamed of snowdrifts
which were the white spores of death caps
drifting across roads
But it was ghosts
who had sent them there
once human, and aware
And nobody breathed
until the dream sun
shone and did melt them.

# 3. Lore

"There was a large mushroom growing near her, about the same height as herself; and, when she had looked under it, and on both sides of it, and behind it, it occurred to her that she might as well look and see what was on the top of it."

Lewis Carroll, *Alice in Wonderland*

Mushrooms spring up in the night as if from nowhere, adorning forest and field, and then decay as suddenly. They are oddly shaped and unpredictable. Some are manna and ambrosia, a free and unforced gift of the gods, but some make an eater lose his wits or his life; and there is no simple shortcut to know one kind from the other.

They have been loved and feared for these reasons, as well as for their practical edibility. Once they stood in mysterious worlds of folklore and poetry, spooks and gnomes, good spirits and bad. Now they seize the imagination no less: as symbols of what—the unconscious? love, from our literal reproductive apparatus right on up? the evanescent natural world that now so needs our protection—against ourselves? our own lives and even our planetary existence, clearly in new jeopardy? They are something more than free groceries in the woods.

Dubbed "lower plants" in reference to the ancient ladder of excellence whose upper rung (in the created world) human beings occupy, mushrooms have been considered poor cousins of the "higher" plants, and have been much less studied. But the 1979 *American Men and Women of Science* lists 261 mycologists, a respectable showing among the biological sciences and more than we expected to find. Many biologists today dignify mushrooms as a distinct Kingdom of Fungi, alongside those of Animals, Plants, and certain microorganisms. New mushroom books are coming out, clubs are growing, and information is being accumulated at an unprecedented rate.

The mushrooms that grow around the northern hemisphere have been regarded in quite various ways by Anglo-Americans, western and central Europeans, Russians and Chinese. Of all these the

Anglo-American culture has taken the least interest in mushrooms and found them least attractive. Here is a sampling of attitudes and lore from a few of these countries.

## England and Anglo-America

The first thing to notice in our own English-language culture is the names. Birch, gentian, maidenhair fern, pocket gopher, ladybug, White-crowned Sparrow--lots of other living things have good-sounding common English names; but few fungi do. We do have Meadow Mushroom, Shaggy Mane, Man-On-Horseback; but most of the prime eating species, conspicuous and widespread though they are, either have no ordinary name or have only a recently coined and unconvincing one. Gatherers use the awkward Latin words—*Boletus edulis, Lactarius deliciosus, Suillus brevipes.*

The lack of common names is a barrier to knowing mushrooms. To have only scientific names suggests that the subject is so difficult, technical and abstruse as to be beyond the capabilities of persons unwilling to spend years qualifying themselves as experts.

The main reactions to mushrooms in the Anglo-American world have been ignore, abhor. It is possible growing up to spend a great deal of time outdoors and love nature, even to frogs, toads and insects, yet step over the plentiful and handsome mushrooms as if they belonged to a different world; hardly dare touch them, even.

The word "toadstool" collects a number of these feelings. The toad is supposed to be a poisonous witch-creature, and the stool he sits on--or worse still, his excrement--is a kind of quintessential low-scoring substance in the universe.

"The grieslie Todestoole growne there mought I se
And loathed Paddocks lording on the same,"
wrote Spenser in his *Shepheardes Calender.* "Paddock" is toad ("paddock-stool" is another word for toadstool) and this is an ill omen, in springtime, of coming winter and death.

"Mushroom" is often contrasted with "toadstool"--the one edible, the other poisonous. "How do you tell a mushroom from a toadstool?" is the unanswerable question asked by people who have never gathered mushrooms. Both words have been used in the sense of "a despised newcomer". Samuel Johnson's *Dictionary* in

1773 defined mushroom in part as "an upstart; a wretch risen from the dunghill; a director of a company."

When he was twenty-seven and living in mushroom-loving Italy, Shelley wrote a long poem "The Sensitive Plant" in which mushrooms are the image of a natural life so perverted as to be deathlike. A most sensuous flower garden is tended by a Lady, virginal, but who awakens in the mornings blushing from the visits of male spirits. Some lower beings are tolerated here--fireflies, bees, moths--but others are not: she compassionately carries out and dumps in the woods

"...things of obscene and unlovely forms."

Well, Sir, the Lady dies, just like that, at the end of Part Second. The garden undergoes a metamorphosis from Eden toward...the world we all live in, presumably. Leaves drop, winds rise. "All loathliest weeds began to grow"--nettle, thistle, dock, henbane, hemlock--disorderly, unloving, but previously kept away by the Lady's presence. At last,

"agaries, and fungi, with mildew and mould,
Started like mist from the wet ground cold;
Pale, fleshy, as if the decaying dead
With a spirit of growth had been animated!"

"Spawn, weeds, and filth," cries the poet, "a leprous scum...."
Winter comes, then a spring which is a mockery of the earlier summer when the Lady had lived still.

"...mandrakes, and toadstools, and docks, and darnels,
Rose like the dead from their ruined charnels."

All this innocent vegetation made to bear such a load of sin and wickedness! What a contrast with the sweet and sexy flowers of the original "undefiled Paradise" which had been dwelt on in stanza after stanza. Shelley's whole scheme of things is not altogether clear, but it is quite plain where mushrooms stood in it. Let it be a lesson! the Lady should have allowed the "obscene" bugs and things to share her garden.

Mushrooms also figure in Browning's very long poem "Paracelsus." The Faust-like German physician, a contemporary of Martin Luther, scorned old books and treatises and sought to learn direct from nature. The poem begins with his vision of mystic knowledge and power. Although he learns much about curing sickness, he later feels he has failed in his main quest because he has made the fatal error of neglecting human love. Pride, arrogance, stormy temperament, immoderate language topple his originally splendid ambitions, and he is demeaned by the shallow satisfaction he discovers in being a celebrity. Perhaps he pushes it a bit, with quackery and trickery. Anyhow, here is where the mushrooms come in:

> "I hate no longer
> A host of petty vile delights, undreamed of
> Or spurned before; such now supply the place
> Of my dead aims: as in the autumn woods
> Where tall trees used to flourish, from their roots
> Springs up a fungous brood sickly and pale
> Chill mushrooms colored like a corpse's cheek."

Again the poor mushrooms are burdened with meanings of death, failure, disgusting little ignoble lusts.

In North America this idea of mushrooms was domesticated with the English language and much else. Even an essential nature person like Thoreau (according to John Cage) evidently never ate a mushroom. WASPs were not greatly impressed when starting in the late 19th century people from central and southern Europe, from cultures rich in the love and lore of mushrooms, came to America and began harvesting the woods and fields--after all, they were here to learn, not teach, in the great melting pot.

Sylvia Plath in her creepy poem "Mushrooms" *(The Colossus)* has them "Overnight, very/Whitely, discreetly,/Very quietly" sneaking up like a nightmare to take over everything--

> "We shall by morning
> Inherit the earth.
> Our foot's in the door."

To this day English and American mushroom books give high prominence to the dangers of poisoning, the difficulties of properly recognizing mushrooms unless one is a certified expert with microscopes, chemicals and a university degree, and they downplay the joys and rewards. An English book labels mushroom after mushroom as "worthless" even if not actively toxic--an unnecessarily strenuous throw-away term. Alexander H. Smith, long an outstanding American mycologist, has taken a dim view of mushroom-eating and is himself allergic to most species. A typical warning, in a book about mushroom poisoning, takes the form: "Don't play Russian roulette with your life for a few cents worth of vegetables."

Funny they should mention the Russians, to whom, as we shall see, mushroom eating is part of the whole purpose and meaning of life itself.

Now, this has been the dominant view, but not the only one. All along mushrooms have been gathered and enjoyed by English-surnamed country people both here and in Britain. A certain number of proverbs, recipes and definitions attest to this. For instance,

> "When the moon is at the full
> Mushrooms you may freely pull,
> But when the moon is on the wane
> Wait ere you think to pluck again,"

is reported from 19th century Essex. Lewis Carroll had Alice control her size by nibbles of a friendly mushroom. And Charles McIlvaine in introducing his *One Thousand American Fungi* at the turn of the century predicts that "Some day the delights of a mushroom hunt along lush pastures and rich woodlands will take the rank of the gentlest craft among those of hunting, and may perchance find its own Izaak Walton." Gary Snyder in his poem "The Wild Mushroom" *(Turtle Island)* celebrates mushrooming--here's a stanza:

> "We set out in the forest
> To seek the wild mushroom
> In shapes diverse and colorful
> Shining through the woodland gloom"

--made as if to be sung to guitar, a kind of tongue-in-cheek ballad;

anyhow it says what it says, and it's not the negative old view of mushrooms.

## Russia

For the most complete contrast possible, let us turn to the Soviet Union. If any nation on the planet is devoted to mushrooming, it is the Russians. "In the fall, it approaches a national craze...At peak season, competition gets so keen that groups organize expeditions, rent buses through their factories on Friday, spend the night in the bus on a country road, catnapping or warming themselves with tea or vodka so that at daybreak they can be the first to get a crack at tender new mushrooms" (H. Smith). "There are surely not many people...whose heartbeat does not quicken with joy at the tidings of the appearance of mushrooms in the forest," says a popular manual. "The Mushroomer can and must be put in the glorious tribe of hunters and fishermen. He too, like the hunter and the fisherman, passes many hours alone with nature. He walks over tens or even hundreds of kilometers of forest trails in search of mushrooms. And in this passion the hunter of mushrooms is not afraid of rain, nor frost, nor heat, nor wind."

"Just rise early in the morning," is the advice to one who wishes to join the "glorious company of mushroomers", "take your basket and go to the forest.... The crimson sun, just beginning to show itself above the horizon, has lavishly gilded the treetops, splashing millions of golden sparks over the emerald green foliage. The forest, washed with the fresh morning dew, still threaded with predawn mist, has fallen silent and beckons you into the half-darkness of the fragrant spruce, into the bright groves of lovely white-trunked birch and slender-footed aspen. It has prepared many a surprise over the previous night. Hasten over the grass, wet from the night, sniff the air--there is a special scent, the scent of mushrooms. Under each tree, each bush, a lucky surprise may await you" (Andrest). This was written in the twentieth century, but it is the old fairyland forest of the Russian folk tale, where the cottage on a hen's foot turns its door to face you if you have the right words, birds and animals talk, and the witch-like old lady knows all. "Unnoticed I follow my grandmother into the thick of the forest; silently she floats among the mighty tree tunks and, as if diving, bobs down toward the pine-needle covered ground...By the scent of the plants she knew what mushrooms ought to be in one or another place, and often quizzed me: 'What tree does a ryzhik like? How do you tell a good syroyezhka from a poisonous one? Which mushroom likes ferns?'" (Gorkiy, in Andrest). The little book even manages to imply that it is unpatriotic not to love mushrooms.

There are about 300 species of edible mushroom in the USSR but only a few are commonly used as food. The best edibles are considered to be *Boletus edulis,* two species of *Leccinum, Suillus,* some *Lactarius,* and the Honey Mushroom. Various Russulas and *Lactarius deliciosus* come next. With very few changes, the list would serve a mushroom hunter in the United States as well. Mushrooms are far more internationally-minded than either Russians or Americans.

The Russian language has words for many kinds of mushroom, and poeple are not obliged to use awkward coinages or Latin names as English speakers must.

What could be the reasons for the huge contrast between Anglo-American and Russian feelings about mushrooms? It must have to do with forests. England is, and has been for 400 years, mainly open country broken by hedgerows and only occasional parklike woodlands. British "nature" is a calm, neat, tended garden--not dramatic wilderness, high mountains, deep woods. The disorderly unpredictable something that fungi substantiate, and their association with the untamed outland, is antipathetic to many British souls.

In Russia on the other hand, the forest, where most of the mushrooms grow, is the essence of the wild and the limitless. The Siberian forests are approximated only by those of the Amazon Valley in their enormous size, and the Russians are in love with them. Many would not trade the forests of central Russia for all the dazzling south: "the mushroom-scented air of the birch groves is far dearer than the fragrance of the magnolia" (Paustovskiy, in Andrest). The typical Russian garden is an untended fragment of wilderness, an appealing contrast to over-crowded, over-supervised life outside.

North Americans are in between--much of our culture including ideas about mushrooms came from England, but we have immense forests like Russia, which draw our imaginations.

"One of my wishes is that those dark trees,
So old and firm they scarcely show the breeze,
Were not, as 'twere, the merest mask of gloom,
But stretched away unto the edge of doom."
Robert Frost

The unknown and limitless woods have a pull that many Americans have felt but which would be atypical for an English person. Although the popularity of mushrooming is increasing in England too, we would guess that it is Americans, over the next decades, who will take to mushrooms more in the Russian style: though the society is so different, and in many ways more amiable, than the Soviet Union, there are still sufficient constraints and frustrations to make us long for wildness, and still forests enough to roam in.

## Germany

The Germans too are lovers and gatherers of the wild mushroom, if not with quite the high pitch of devotion the Russians have. Germany's traditional romantic love of nature--or Nature, because She often takes the form of a quite concrete figure--embraces mountains and forests, and rejoices in their freedom and truth as compared with human settlements. "Im Walde bin ich König," "in the woods I am king," says the folk song; "I'll build you a little hut of evergreen twigs...and never go back down to the gray village."

Every fall Germans stream out of the cities to gather mushrooms in the woods. As the proverb says, "Man muss die Pilze suchen, wenn sie im Walde wachsen", (you must hunt for mushrooms when they're growing in the forest)--strike while the iron is hot. But other mushroom proverbs carry warnings:"Je giftiger der Pilz, je schöner die Kappe" (the more poisonous the mushroom the more beautiful its cap); "Wer alle Pilze brät, kann sich leicht vergiften" (he who cooks all mushrooms can easily poison himself). And

"What's born a mushroom will never be a cedar" (you can't make a silk purse out of a sow's ear). The proverbs attest to the culture's long-standing interest. In the Low German of the North, there has been an association (as in English) with toads: "Poggenstool" (toadstool), with the "stool" being either seat or excrement.

There is a double feeling, with the maternal Nature figure of the mushroomy forests stern and guilt-inspiring as well as loving and nurturing. Mushroom hunting can have a misanthropic, even masochistic, shading.

> "Of his own accord [the friend of nature] will seek untouched nature and quiet where they are still soonest to be found, in wood and field. Neither beside lakes and rivers, nor even in the high mountains far from cities, will he shout, or leave behind trash and destruction as sorry testaments of his presence--not only because mushrooms, friendly forest goblins, hide themselves from all uproar, but because the true friend of nature feels that beautiful Nature herself flees from any such shameless person . . . .

> "All around the big cities and everywhere that the irreverent city-dweller has shown his artificiality, the forest is impoverished and the fields turned into sandy steppes. For Nature avenges herself, silently but pitilessly, in catastrophes and hunger upon the heirs of the transgressors." (Zeitlmayr)

Man is also a parasite, just like mushrooms: both "suck life from the substance of others, who are thereby destroyed, and kill that they themselves may live." The mushroom hunter is admonished to appreciate *all* Nature has to show, not just the birds and flowers. No mushroom--or any other natural thing--is "worthless":

> "If [the mushroom hunter] at once accustoms himself to walking slowly and not just staring dull-wittedly at the ground, but also raising his glance freely and enjoying all the beauty of wood and field, then he discovers mushrooms also on the trees . . . and mosses, lichens . . . lower fungi. Also he sees then all kinds of creeping, crawling, running, flying life of small animals, of whose co-existence he had hitherto only a vague sense. One sees bees and ants, beetles and spiders, snails and worms, dragonflies and

butterflies, frogs and snakes, lizards and salamanders, mice and moles, birds and game."

It is tempting to go on and speculate about German national character on the basis of this material, but let us just add that this people's old and deep-rooted love of mushrooms is borne out by their language. The common mushrooms, which are again almost all the same as the best ones we find in America, are blessed with a plethora of folk-names, quite the contrary of English where some of the finest suffer the embarrassment of having no common name at all. Thus the Honey Mushroom, *Armillariella mellea,* commonly "Hallimasch", is also Honigpilz, Honigringling, Heckenschwamm, Halamarsch, Wenzelspilz, Michaelischwamm ("Michael's Mushroom", presumably in honor of Michaelmas, September 29, about the season for these mushrooms); Hohlmütze, Stuakschwammala, Stubbling, and Schulmeister (Zeitlmayr). The mushroom has succeeded in getting the Germans' attention.

## Ancient Rome and Italy

The ancient Romans thought wild mushrooms one of the finest gourmet foods, but they were wary of the risks, with some reason because poisonous mushrooms had a reputation as murder weapons. The favorite edible was *Amanita caesarea,* which is found in the woods of eastern North America and in the mountains of the southwest, as well as in southern Europe. The names that the Romans used are not the same used in scentific Latin today. *A. caesarea* was "boletus" in old Latin, and our *Boletus,* in a different family, was their "suillus" or pig-mushroom, and not the genus we call *Suillus* now.

Pliny the Elder's *Natural History* contains a long section in which he describes the good and bad Amanitas, with their universal veil, quite clearly, along with other mushrooms (Book XXII, 46.92 ff). "How chancy a matter it is to test these deadly plants!" he says.

> "If a boot nail, a piece of rusty iron, or a rotten rag was near when the mushroom started to grow, it at once absorbs and turns into poison all the moisture and flavour from this

foreign substance....If the hole of a serpent has been near the mushroom, or should a serpent have breathed on it as it first opened, its kinship to poisons makes it capable of absorbing the venom."

Such ideas deceived mushroom-pickers for centuries and to a degree still do.

The Italians today enjoy mushrooms as their Roman ancestors did, gathering and eating *Amanita caesarea* and a limited number of other species, most of which also grow in America. Many edible mushrooms go to waste because of inadequate knowledge, "and represent a not negligible loss to the economy of the country" (Viola). Those that are well known are gathered in quantity, and as in Germany the language and its dialects are full of names for them. Thus the princely *Boletus edulis* (and immediate congeners) is in standard Italian "porcino" (referring to pigs, like the Latin Suillus), but also (take a deep breath): boleto dei pini, ceppatello buono; funzo neigro or funzo gnaco (Liguria); nivarieu, farieu, duret, caplet, bolè (Piemonte); ferée, fung ferée, levrin, legorsela, vairoeu, brisott, none, boler dur, luco, cucola, frer (Lombardia); boléo, sbrisà, sbrisato, brisa (Veneto); moreccio (Toscana); cunzella, blué nigher, durun, fonz curpadel, nigrou, blisgon (Emilia); lilli, munito (Napoletano); vavasu, lardito (Calabria); cardulino gragu (Sardegna). It's enough to make an American jealous, with our clumsy "King Bolete" which nobody says, or "cep" after the French name for this mushroom. Perhaps we should just borrow one or another Italian name--surely they don't need them all--vavasu, say, or fonz curpadel.

An Italian mushroom rhapsody has its own special tone.

"Each kind [of mushroom] has a particular odor or perfume which, blended with that of the ferns and mosses, produces that delicate inebriating scent of the forest, so fragrant, so pleasing, so characteristic. After being immersed for a short time in this storyland atmosphere, new thoughts come to the surface of the mind, memories of things and events from forgotten years. Little by little, many of our superstructures accumulated in civil, or pseudocivil, life fall away and vanish, and something ancestral, primitive, resurfaces in us. A sensation of freedom, or purity, invades us and lightens us: all that surrounds us is ours, belongs to us, not through dominating or being dominated but because we feel we are part of it, are living the same life.'' (Viola)

What a difference from the Anglo-American way of looking at the risks of poison from mushrooms:

"In reading of the *Amanita phalloides* and of the frightful, though rare, danger this mushroom presents, and of the awful death it brings, one will wonder why nature is ever so cruel (cruel toward man), and in such seemingly innocent, almost gay, forms.... The important thing is that, beyond all contradiction and all distress, there can always be discerned a marvellous and continuing universal harmony.... Of that harmony the tender little mushrooms that can be gathered, like the immense incandescent unapproachable stars, would be, more than symbols, symptoms: living peepholes in a great door that is still too opaque for us." (Soldati, in Viola)

This apparently doesn't sound silly in Italian, and after all why should it in English? Maybe someday effusions of happy emotion toward mushrooms won't seem inappropriate any more than toward roses, nightingales, or skylarks.

## France, Spain

"France," says a French reference work (*Encyclopaedia Universalis,* 1968), "occupies a select place among mycophagous [i.e., mushroom-eating] nations." They do have a reputation. Dr. Paul Ramain distinguished 100 tastes and 275 aromas in wild mushrooms (*Mycogastronomie,* 1953, cited in Viola). More recently Marcel V. Locquin explains the following terms:

"--A 'pot hunter' is someone who knows and eats only two or three species;
--a 'mycophage' is one who has enjoyed a score of species;
--a 'mycogastronome' is one who has enjoyed more than one hundred and selects from them those he likes best."

"The author of this book," he goes on, "has tasted more than 700. He has retained about 300 as possibles, and a hundred or so for his current consumption." (*Mycologie du Goût*). French mushroom cookery is like traditional French cuisine in general, with its advantages and its limitations. It is rich, complex, and subtle, and uses lots of butter, cream, egg yolk, wine, and paprika.

Some of the favorites of French gatherers are commonly found in North America also: *Boletus edulis* (cèpe), *Lactarius deliciosus* (lactaire délicieux), morel (morille), Shaggy Mane (coprin chevelu), and Chanterelle (the French word, also girole).

A Spanish proverbial saying that has a plain charm is "solo como un hongo"--as alone as a mushroom, meaning without friends and without dealings with other people. Thus Wordsworth might have said, had he been a Spaniard: I wandered lonely as a mushroom.

## China

For at least two thousand years a certain Magic Mushroom called lingzhi (or shenzhi) played the part of a Philosopher's Stone or Holy Grail in China. The lingzhi was a good omen, but more than that it was believed capable of bringing about a complete unfolding of human life: immortality, or more subtly, the full deployment of knowledge and capacity that was the Daoist preoccupation.

Pictures and descriptions point to the genus *Ganoderma* and perhaps to *G. lucidum,* a tough and inedible bracket fungus that occurs also in North America. It is beautifully streaked and shiny as if varnished, and itself long-lived and resistant to rot. It it not psychotropic (which was an interesting possibility), and has no other drug properties. But the literal identification is not the point, any more than it matters greatly whether the Holy Grail was a pitcher or a bowl.

Late in the third century B.C. the First Emperor of Chin, the ferocious unifier of China, sent out an expedition into the Eastern Sea to find him this "spirit-nourishing" mushroom so he could be immortal. Luckily the voyage was a failure.

Around 300 A.D. the philosopher Ge Hong wrote at length on

the various classes of zhi mushrooms of interest to seekers of the Dao. He divided them into five groups: Stone, Wood, Herb, Flesh, and Fungus Mushrooms, and reports the mythic qualities of each. Again the connections with any literal mushrooms are tenuous and secondary.

On Stone mushrooms, for example, he writes:

"Mushrooms shaped like stones grow on out-of-the-way seacoasts, famous mountains, and the shores of rocky islets where stones are heaped. They have fleshly shapes, some with head, tail and four limbs, just like a living creature. They attach to boulders and prefer high places and steep cliffs. Some reject firm attachment and cling to each other instead. There are red ones like coral; white ones like sliced fat; black ones like lacquer; green ones like kingfisher feathers; yellow ones like purple gold. All are glossy and translucent like firm ice. On a dark night their light can be seen from a distance of three hundred paces. Large ones weigh upwards of ten pounds, small ones three or four.

"Without having fasted a long time, keeping the utmost singleness of mind, and carrying Laozi's Magic Treasure Five Charms for Entering the Mountains, it is impossible to have sight of these kinds. Upon seeing any mushroom, first place upon it the Charm for Opening the Mountain and Driving Away Harm, so that it will not hide again, change itself, and be gone."

One must select an auspicious day and perform the correct ceremonies before gathering, proceeding always in the direction of the sun, using the "gait of Yu," and holding the breath. The mushroom is ground up with 36,000 strokes of the mortar. One pound, taken in measured doses, gives a longevity of a thousand years; ten pounds, ten thousand years.

Yu was a sage-emperor at the dawn of history who injured a leg in his labors to control a great flood. As a result he walked with a hobbling gait, adopted later by shamans and magicians: as Ge Hong describes it, it means a step forward with the right foot, one with the left, then the right brought up even with the left; repeated beginning with the left foot. It is a dance stylization of what a mushroomer actually does in the woods, starting and stopping, going this way and that, as if indeed lame.

One of the Stone Mushrooms, after a pound of it has been consumed, makes the body shine like the moon so that one can see in the dark.

One of the Wood category, the weixizhi ("Dignified and Delightful Mushroom"), grows where pine pitch has soaked into the ground for a thousand years and changed into the underground fuling fungus; on the surface above this, after another ten thousand years, the weixizhi arises as a small tree of lotus shape. "It glows in the dark. If you grasp it it is very slippery. It will not burn. If you carry it on you it will ward off weapons. Fasten it to a chicken and put that chicken in a cage with a dozen others, walk off twelve paces and shoot twelve arrows--the other chickens will all be wounded, but not the one with the weixizhi." It too gives long life, measured in the thousands of years.

All these, and the hundreds of other magic mushrooms, are rare and difficult to find. The common run of would-be Daoists, "whose mind is not pure and concentrated, whose conduct is unclean, whose virtue is shallow and who do not know the arts of entering the mountains, will not know their shapes even with a picture, and will never obtain them. Large or small, mountains have their ghosts and spirits and if the ghosts and spirits will not give mushrooms to a person, he will not see a mushroom even if he steps on top of it."

Ge Hong wanted an alchemy from mushrooms: he wanted the detailed knowledge of a scientist, and the depths of a mystic. Mushrooms are very apt to evoke these twin wishes in their seekers still today.

Mushrooms in China had simple divinatory meanings too. A ninth century writer, Duan Chengshi, remarks that "When housepillars produce mushrooms for no apparent cause, white ones mean mourning; red ones, blood; black ones, bandits; yellow, joy; those shaped like a human face, loss of wealth; like cattle and horses, a long journey; like turtles and snakes, gradual deterioration." The sixteenth century botanist and pharmacologist Li Shizhen

Plate 1

*Lycoperdon perlatum*
Puffball

p 28

p 28

*Calvatia*
Puffball

past edible condition

*Morchella angusticeps*
Black Morel

p 30

Plate 2

*Dentinum repandum*
Hedgehog Mushroom

p 32

*Hydnum imbricatum*
Owl Mushroom

p 32

*Suillus lakei*

p 34

*Suillus brevipes*

p 35

Plate 3

*Suillus granulatus*     p 35

*Boletus edulis*
King Bolete

p 35

*Boletus edulis*
King Bolete

p 35

Plate 4

*Leccinum aurantiacum*

p 35

*Cantharellus cibarius*    p 42
Chanterelle

*Cantharellus cibarius*
Chanterelle

p 42

Plate 5

*Lactarius deliciosus*

p 42

*Pluteus cervinus*
Deer Mushroom

p 42

*Coprinus comatus*
Shaggy Mane

p 43

*Gomphidius glutinosus*
Yellow Foot

p 43

Plate 6

*Agaricus campestris*
Meadow Mushroom

p 43

*Agaricus sylvicola*
Wood Mushroom

p 43

*Pleurotus ostreatus*
Oyster Mushroom

p 43

*Flammulina velutipes*
Winter Mushroom

p 43

Plate 7

p 44    *Armillariella mellea*
Honey Mushroom

*Armillariella mellea*
Honey Mushroom

p 44

*Armillaria ponderosa*
Pine Mushroom

p 44

Plate 8

*Russula decolorans*

p 44

*Tricholoma flavovirens*
Man-on-Horseback

p44

*Amanita pantherina*

p 24

*Amanita muscaria*
Fly Agaric

p 24

cited this passage with approval, but could not hold with the idea that immortality can be gained from mushrooms. "I have always thought that mushrooms are the product of energy (vapors) left over from decomposition," he wrote, "just as people produce wens and tumors. But from ancient times until the present, all have considered them plants of good augury.... That taking doses of them can lead to immortality is certainly erroneous."

Apart from their magical aspects, mushrooms have been a favorite food of the Chinese, who in general eat more foods prepared in more ways than any other people on earth. They enjoy the wild savor of mushrooms, and speak of them as the equal of meat in richness and flavor, using them extensively in vegetarian diets (as did monks in Europe and Russia). One old poem (whose authorship we did not discover) could refer to the Deer Mushroom or the Winter Mushroom:

"What luck! they spring up on a dead tree,
And go to be blended in our mouths.
They have the authentic taste of woods and mountains,
Difficult to articulate to ordinary people.
A trace of fragrance wafts from the jadelike caps,
A sense of vitality imbues the elegant stalks.
I gobble them in such quantity that I'm ashamed,
And in return can offer only a poem."

The twelfth century poet Yang Wanli wrote this celebration of a mushroom that might perhaps be the Chanterelle:

"As soon as it rains, in the wild mountains,
    Streams run faster,
Carrying along buds of the cassia,
    And pine cones.
The rich soil is soft and warm,
    All juices soak in,
And troops of mushrooms spring up
    In the steamy vapors.
Wearing fallen leaves like caps,
    They suddenly uprise,
Push off their fallen leaves,

> Hundreds of them.
> Waxy caps, brownish yellow,
>> Glisten in the wet.
> Tender stalks, crisp and delicate,
>> Snap in the hand.
> Color like the foot of a goose,
>> Fragrance like honey;
> Texture smooth as the new Water-shield,
>> Never a hint of harshness.
> The caps are not as good as the stems,
>> Buttons are better yet.
> The flavor lingers on the palate,
>> Finer than musk.
> Lamb-and-cabbage, chicken cooked in paper
>> Must yield pride of place.
> Dinners of jade elegance, the magic zhi-fungus
>> Cannot be compared.
> Made into soup, they will not keep
>> For a single day.
> Salt and ferment them, though, and you
>> May fill your larder.''

(Water-shield is an edible water plant of mushroom shape.)

More kinds of mushroom are probably cooked and eaten in China than anywhere else, including all but two or three of the species discussed in this book.

\* \* \*

These are just a few of the ideas and emotions people have come up with. Mushrooms in art have their own story; mushrooms in music--John Cage said, ''I have come to the conclusion that much can be learned about music by devoting oneself to the mushroom'' (*Silence*). The role of mushrooms in religions has been told by Gordon and Valentina Wasson. Mushrooms are not only wild things in their fine carelessness, but valued human objects of culture. The forest is a museum.

*Tricholoma flavovirens*

In the forest above this valley
if you will look through the tall pines
here and there Man-On-Horseback
raises a yellow-green cap,
tan-centered, barely beyond
his sandy nest, hardly
clear of the litter, where
most of his dry sporedust
falls back. Now and then,
though, a white spore climbs
on air, and sails
halfway around the world
to, perhaps (odds are against),
high ground, pines, sandy nest.

# 4. Gastronomy

When lovers of fine food are asked to name their most memorable meal they almost invariably tell about Aunt May's homemade preserves on homemade bread with freshly churned sweet butter; or about the day, driving across the French countryside, that the car collapsed and the farmer's wife served salad of garden vegetables, and milk-fed veal with a wild mushroom sauce; or something similar.

The finest culinary experiences are most often something homemade, home grown, home concocted. Restaurants can only try to emulate this, by sending hither and yon for the freshest, best ingredients.

It is curious, therefore, that the word gourmet has become so overused and misunderstood. To many people it seems now to mean expensive, elaborate and rich. A "gourmet menu" usually turns out to be a horrible list of ten major dishes that leaves the diner uncomfortably overstuffed and dieting for a week afterwards.

In fact, truly fine food is a matter of the quality of ingredients, the skill and dedication of the cook, and an array of dishes that is at once sophisticated and restrained.

Mushroomers are often quick to realize this. A small plate of lightly seasoned, sauteed, fresh, young *Boletus edulis* served with thin, crisp wafers, is as much of a gourmet snack as can be imagined. If a famous French chef dropped in to a mushroomer's house for dinner and was served cream of chanterelle soup made from mushrooms picked an hour before he would go into gourmet shock. And, if he was worth his salt, would spend years trying to reproduce the recipe in his restaurant.

Mushroom cookery, like mushrooms, is worldwide. All major cuisines boast dozens of recipes featuring mushrooms, and as any dedicated cook knows there is no end to the delights that can come out of the mushroomer's kitchen. This chapter contains a broad range of recipes from many parts of the planet. Some are simple, most are moderately easy, a few are rather complicated. Every effort has been made to use ingredients that are easily obtained, even in the outback where the hunter tends to hang around, but trips to city specialty food stores will be necessary to prepare a few of the recipes.

One of the most unique aspects of the many unique aspects of mushrooming is a culinary one: mushrooms are a powerful flavor enhancer. Almost anything tastes better with mushrooms. Needless to say, this is an outstanding feature of fungus cookery. A fine recipe may call for a relatively small proportion of mushrooms, and yet it would be quite flat and ordinary if they were omitted.

## Food Value

"It is noticeable," says a Russian writer on mushrooms (Andrest), "that individuals who constantly consume mushrooms feel good as a rule. This means that with mushrooms, a human being gets a sufficient quantity of those substances necessary to the organism." But "fungi in general do not have a high nutritive value" says a presumably British author in Hong Kong (Griffiths). The fact is that mushrooms as food carry a high degree of mystique, and how nutritious they are thought to be depends on where you are.

Apart from actual food value, their scents, flavors and amino acids are as stimulating to the appetite and digestion as their woodsy mysterious associations are to the imagination. But mushrooms really do have much value as food. Although they vary greatly among species (and between individual mushrooms as well) they compare well with vegetables and meats in their offerings of food energy, protein, vitamins and minerals (see Tables I, II and III). They are generally high in the B vitamins, calcium, phosphorus, and iron, among nutrients that tend to be deficient in American diets. Some nitrogenous material, counted with the protein, is tied up as fungin, identical to the chitin which forms the horny exoskeleton of insects. But even allowing for some indigestibility and some limitations imposed by amino acid imbalances, the protein quality of many mushrooms is excellent, ranking with some meats and above such vegetable protein sources as soy beans and peanuts.

## TABLE I

### Nutritive Value of Proteins

FAO Amino Acid Score for Quality of Protein (Egg = 100)

| | |
|---|---|
| 100 | egg, pork |
| 98 | beef, chicken |
| 91 | milk |
| 36-90 | *Agaricus bisporus* |
| 88 | *Leccinum scabrum* |
| 76 | *Dentinum repandum* |
| 68 | *Cantharellus cibarius* |
| 57-60 | *Pleurotus ostreatus* |
| 53 | peanuts |
| 40 | *Morchella angusticeps* |
| 37 | *Boletus edulis* |
| 28 | Spinach |
| 23 | Soybeans |

Crisan and Sands, in Chang (ed.).

All across the old world, the flavor and the protein value of mushrooms have long made them favorite ingredients of vegetarian and fasting diets, taking the place of meat. They are also valuable additions to a meager fare of grains or potatoes for country people. As far as that goes, squirrels, chipmunks, deer, elk, grouse and other wildlife often feed on mushrooms. (That animals, birds or insects eat a mushroom is NOT a sign that it is safe for human beings.)

There are many other valuable food substances in mushrooms. One hundred grams of the Honey Mushroom, *Armillariella mellea,* contains the daily requirement of zinc and copper, and some mushrooms contain vitamin D. *Boletus edulis, Lactarius deliciosus* and Chanterelles contain vitamin A. Mushrooms contain lecithin, and among the carbohydrates, glycogen and a special mushroom sugar, mycose. One simple reason nutritional scores appear low in

## TABLE II
## Vitamins and Minerals
## mg per 100 g

| | Thiamin | Ribo-flavin | Niacin | Ascorbic Acid | Calcium | Phos-phorus | Iron | Sodium | Potas-sium |
|---|---|---|---|---|---|---|---|---|---|
| *Flammulina velutipes* dry weight | 6.1 | 5.2 | 106.5 | 46.3 | 19 | 278 | 11.1 | 278 | 2981 |
| *Pleurotus ostreatus* dry weight | 4.8 | 4.7 | 108.7 | 0 | 33 | 1348 | 15.2 | 837 | 3793 |
| *Agaricus bisporus* dry weight | 8.9 | 3.7 | 42.5 | 26.5 | 71 | 912 | 8.8 | 106 | 2850 |
| Whole wheat bread made with water (unenriched) | .30 | .10 | 2.8 | T | 84 | 254 | 2.3 | 530 | 356 |
| Beef (cooked, lean hamburger) | .09 | .23 | 6.0 | — | 12 | 230 | 3.5 | 48 | 558 |
| Spinach, boiled and drained | .07 | .14 | .5 | 28 | 93 | 38 | 2.2 | 50 | 324 |
| Dandelion greens, boiled and drained | .13 | .16 | — | 18 | 140 | 42 | 1.8 | 44 | 232 |

Chang (ed); U.S. Department of Agriculture Handbook No. 8.
Note: samples vary; thus other figures for iron in *Agaricus bisporus* are 0.2, 19.0, 128! (Chang, ed.).

some tables is the high water content of mushrooms, around 90%. In Tables II and III dry weights of mushrooms (or in one case salted mushrooms) are used.

Mushrooms should be gathered in clean places, picked and transported with care and washed as little as possible. As food, they are like other food and only the freshest and best should be eaten. They do not keep well in a fresh state, whether in the woods or at home, particularly in warm weather.

Some mushroomers eat a number of species raw, but we do not recommend this. Slicing thin and sauteeing lightly in a little butter with a dash of salt seems to be the best way to study a new species. This method brings out the flavor and texture most dramatically. Begin sampling just moments after it has been combined with the

## TABLE III

### Protein, Fat, Carbohydrate and Calories

|  | Protein g/100 g | Fat g/100 g | Carbohydrate g/100 g | Calories per 100 g |
|---|---|---|---|---|
| *Boletus edulis,* dried | 33.0 | 13.6 | 26.3 | 224.2 |
| Powdered *Agaricus* (prob. *bisporus*) | 45.5 | 3.8 | 20.9 | 192.0 |
| *Lactarius deliciosus,* preserved with salt | 21.85 | 3.75 | 47.75 | 183.7 |
| Whole wheat bread made with water (unenriched) | 9.1 | 2.6 | 49.3 | 241 |
| Beef (cooked, lean hamburger) | 27.4 | 11.3 | 0 | 219 |
| Spinach, boiled and drained | 3.0 | 0.3 | 3.6 | 23 |
| Dandelion greens, boiled and drained | 2.0 | 0.6 | 6.4 | 33 |

Andrest; U.S. Department of Agriculture Handbook No. 8.

butter, and continue until it is crisp. Individual tastes vary of course, but most will agree that there is an optimum point of flavor for each mushroom (we have given our opinion with the discussion of the individual mushrooms). Experiment with various herbs and spices. After discovering their individual characteristics it can be determined for what sort of more elaborate recipe they would be best suited. Some, like *Lactarius deliciosus,* stay at their best simply sauteed. Others, like *Boletus edulis,* seem to be best practically every way.

Fresh mushrooms can always be substituted for frozen. Fresh can also be substituted for dried unless the recipe calls for the rehydrating mushroom water (as in many of the Chinese dishes).

In most of the recipes one mushroom is suggested. Particularly for the person new to wild mushrooms, it should be pointed out that one variety often can be used in place of another. Attention should be paid to strength of flavor and texture, of course, but since mushrooms have seasons and are so capricious, cooks substitute when necessary.

## Freezing

Clean and thinly slice the mushrooms. In the case of the bigger boletes or the slimy topped *Suillus* prepare as you would for sauteeing by removing the pores or tubes and/or peeling the top. Boil no more than one pound in one gallon of water for 2 minutes. Ad 1/2 minute to blanching time for each 2,000 feet above sea level. Run under cold water to stop the cooking. When completely cooled freeze in bags like other vegetables or spread on cookie sheets covered with waxed paper and bag when frozen.

Or, freeze any but the largest whole. Blanch small mushrooms 2 minutes and big mushrooms 4 minutes (adding 1/2 minute for each additional 2,000 feet above sea level). These are less convenient than if sliced before freezing.

## Drying

Clean and thinly slice the mushrooms. In the case of the bigger boletes or *Suillus* prepare as you would for sauteeing by removing the pores or tubes and/or peeling the top. Spread them out on nylon netting or screens in a clean dry spot. Place in jars when completely dried out--it takes about a week in a dry climate. They should almost crumble to the touch. If not dry enough they will not keep well.

Mushrooms can also be dried whole in the same manner, but it takes a little longer. Other methods include food dehydrators or stringing on sturdy thread.

To rehydrate any mushroom simply place in very hot water and leave until soft. Sliced they take about 15 minutes, whole somewhat longer. Keep in mind that mushroom water has a lot of flavor and is good as a broth or stock. Dried mushrooms have a somewhat different taste and texture from fresh or frozen ones.

In almost every case, we prefer frozen mushrooms to dried.

# Appetizers

### Sauteed *Coprinus* with Parmesan Cheese

*Coprinus comatus* (small ones are best)
finely grated parmesan cheese
whole wheat flour
salt
freshly ground black pepper to taste
lightly beaten egg
butter

Wash the mushrooms but leave whole. Mix together the cheese, flour (about one-half cup of grated cheese to three tablespoons flour), salt and a few grindings of black pepper. Dip mushrooms into the egg and then roll in the cheese and flour mixture. Melt the butter over medium-low heat and saute the mushrooms until golden brown.

*This recipe was suggested by Cheryl Johnson, and is a delicacy that dazzles the finickiest of gourmets.*

### Domestic Mushrooms Stuffed with Wild Mushroom Duxelles

½ ounce dried *Leccinum* (about ½ cup)
6 *Agaricus bisporus* caps, wiped clean
the stems of the *Agaricus bisporus*
1 tablespoon butter
1 scallion, including some green, finely chopped
pinch sage
¼ teaspoon salt
freshly ground black pepper
1 tablespoon port wine
1 tablespoon sour cream
butter

Soak the *Leccinum* in hot water for at least 15 minutes until softened. Drain, discarding the water, and chop fine.

Remove and finely chop the stems of the *Agaricus* and squeeze out the liquid in paper towels.

In a small saucepan or skillet, over low heat, melt 1 tablespoon of the butter. When the foam subsides, add the scallion, mix well and add the sage and chopped mushrooms. Mix and stir in the salt and a few grindings of black pepper. Saute for 3-4 minutes and add the wine. Cook 2 minutes more, stir in the sour cream, heat, stirring constantly and remove from the fire.

Preheat the oven to 350°. Fill the *Agaricus* caps with the chopped mushroom mixture and dot liberally with butter. Place in a buttered baking dish that will hold the caps in one layer and bake in the middle of the oven for 20 minutes.

*Duxelles is one of the most famous of all mushroom dishes and there are practically endless recipes using this mixture. The classic ingredients are mushroom, chopped stems, scallion or shallots, salt, pepper and lemon. From there it is up to the imagination of the cook and the personality of the mushroom. Leccinum, particularly dried, can handle a lot of strong flavors and it comes through admirably in this dish. Its springy texture is very pleasing as duxelles.*

## Stuffed Mushrooms with Spicy Mustard

16 Meadow Mushrooms (all approximately the same size--about 2 "
  in diameter)
1 tablespoon soy sauce
1 small clove garlic, finely chopped
1 tablespoon Dijon, Dusseldorf or other spicy mustard
2 slices cooked bacon, finely chopped
1 tablespoon dry white wine
butter

Preheat the oven to 450°. Remove the mushroom stems, chop fine and squeeze out the liquid in paper towel.

In a small bowl mix the chopped stems, soy sauce, garlic, mustard, bacon and wine thoroughly. Stuff the mushroom caps with the mixture and dot liberally with butter. Place in a buttered baking dish that will hold the caps in one layer and bake in the middle of the oven for 15 minutes, until lightly browned.

**Cheese and Olive Stuffed Mushrooms**

16 Meadow Mushrooms, wiped clean
1 tablespoon butter
1 large scallion, including some green, finely chopped
1 large clove garlic, finely chopped
1 tablespoon parsley, finely chopped
10 pimiento-stuffed green olives, finely chopped
2 tablespoons shelled and salted sunflower seeds
1/8 teaspoon salt
freshly ground black pepper to taste
3 tablespoons parmesan cheese, freshly grated
butter

Preheat the broiler. Remove the mushroom stems, chop fine and squeeze out the liquid in paper towel.

In a small saucepan or skillet, over low heat, melt 1 tablespoon of butter. When the foam subsides, add the scallion and garlic. Saute for 2 to 3 minutes until wilted. Add the mushroom stems, parsley, olives, sunflower seeds, salt and a few grindings of black pepper. Mix thoroughly and cook for 2-3 minutes. Remove from the heat and add the cheese.

Stuff the mushroom caps with the mixture. Place in a shallow dish suitable for broiling and large enough to hold them in one layer. Dot each with a little butter. Broil for approximately three minutes or until the butter is melted and the mushrooms are lightly browned.

## Dried Mushrooms Chinese Style

dried mushrooms (see note below)
peanut oil
fresh ginger root, peeled and chopped fine
garlic, finely chopped
soy sauce

Soak the mushrooms in hot water for at least 15 minutes until softened. Drain, discarding the water.

Heat a 12″ wok or iron skillet over high fire for about 30 seconds. Add a little oil and heat for another 30 seconds. Stirring constantly brown the mushrooms (this dries them out) and then add the garlic and ginger. Mix thoroughly and pour in soy sauce to taste (this reconstitutes the mushrooms again--and gives them a very distinctive taste and texture). Serve as an appetizer or a garnish for vegetables.

*We have tried this recipe with dried* Boletus edulis, Lactarius deliciosus, *puffball and* Leccinum *and it is good to excellent in all cases. It was particularly nice with* Lactarius deliciosus *which, dried, has no taste and poor texture in other methods.*

## Poor Man's Caviar

to serve 4 as appetizer

1 tablespoon oil
1 tablespoon butter
1 small onion, chopped medium fine
1 clove garlic, finely chopped
4 ounces frozen, sliced *Lactarius deliciosus*
¼ cup green pepper, chopped medium fine
1 tomato peeled, seeded and chopped medium fine
½ teaspoon salt
freshly ground black pepper
1 teaspoon lemon juice
dark rye bread

In a 12 " skillet or 2 quart saucepan, over moderate fire, heat the oil and melt the butter. When the foam subsides, saute the onion and garlic for 3 minutes until wilted. Add the mushroom, green pepper, tomato, salt and a few grindings of black pepper. Cook, over low heat, for 5 minutes until the mushroom is defrosted and most of the liquid is gone. Sprinkle on the lemon juice and serve with dark rye bread.

Lactarius deliciosus *seems particularly meaty in this dish. The traditional Russian Poor Man's Caviar is made with eggplant instead of mushrooms. As good as that is, this is better.*

**Three Season Salad**

to serve 2

2 tablespoons peanut oil
1 teaspoon butter
3 ounces sliced, frozen *Boletus edulis*
1 tablespoon soy sauce
3 tablespoons vegetable oil
1 teaspoon sesame seed oil
½ teaspoon Chinese rice vinegar (or mild white wine vinegar)
1 cup dandelion greens, washed and chopped medium fine
1 cup lettuce, washed and chopped medium fine
1 clove garlic, finely chopped

In a small saucepan or skillet, over low fire, heat the oil and melt the butter. When the foam subsides, add the mushroom and cook, stirring frequently, until crisp and golden, about 15-20 minutes. Add the soy sauce, mix well and remove the pan from the heat.

In a salad bowl large enough to hold the ingredients comfortably, combine the 3 tablespoons vegetable oil, sesame seed oil and vinegar. Just before serving thoroughly mix in the dandelion, lettuce, garlic and mushrooms.

*Sesame seed oil can be purchased at oriental food stores. It is not the same as that carried by health food stores.*
*This dish can be served as an appetizer with thin crackers or slices of bread, or as a salad. Spring or late fall dandelion leaves are best because they are less bitter.*

## Soups

### Cream of Chanterelle Soup
to serve 4

2 tablespoons butter
2 tablespoons vegetable oil
½ pound Chanterelles, wiped or washed clean and thinly sliced
3 tablespoons whole wheat flour
1 teaspoon salt
freshly ground black pepper
4 cups vegetable or chicken stock (or 2 bouillon and 4 cups water)
1 cup half-and-half cream
fresh parsley

In a 3-4 quart saucepan over low heat, melt the butter and heat the oil. When the foam subsides add the mushrooms and saute for 2-3 minutes. Stirring constantly, add the flour, salt and a few grindings of black pepper. Stir for a minute or so and then slowly add the stock, stirring vigorously, and bring to a boil. Immediately turn the heat to low and simmer covered for 15 minutes, stirring occasionally. Add the cream and serve garnished with a sprig of fresh parsley.

## Mushroom Soup

to serve 4

2 tablespoons butter
2 tablespoons vegetable oil
1 medium onion, coarsely chopped
2 medium cloves garlic, finely chopped
½ pound mushrooms, wiped or washed clean and thinly sliced
  (see note below)
2 tablespoons whole wheat flour
1 teaspoon salt
freshly ground black pepper
5 cups vegetable or chicken stock (or 3 bouillon cubes and 5 cups
  water)

In a 3-4 quart saucepan over low heat, melt the butter and heat
the oil. When the foam subsides add the onion and garlic and saute
for 5 minutes until translucent. Add the mushrooms and saute for
2-3 minutes longer. Stirring constantly, add the flour, salt and a
few grindings of black pepper. Stir for a minute or so and then
slowly add the stock, stirring vigorously, and bring to a boil. Immediately turn the heat to low and simmer covered for 15 minutes,
stirring occasionally. Serve.

*Frozen or fresh* Boletus edulis, Leccinum, *Meadow Mushroom, etc. can be made into mushroom soup. Dried mushrooms often remain too rubbery to make a good
soup; the only two we have tried and enjoyed are* Boletus edulis *and Meadow
Mushroom.*

**Mulligatawny Soup**

to serve 4-6

4 cups chicken stock
1 whole chicken breast
3 tablespoons vegetable oil
1 onion, finely chopped
½ pound mushrooms, wiped or washed clean and thinly sliced
  (see note below Mushroom Soup).
¼ teaspoon whole cumin seed
1 teaspoon curry powder
2 tablespoons whole wheat flour
4 large stalks celery, washed, and cut crosswise into 1/8 inch widths
1 large tomato, peeled, seeded and coarsely chopped
¼ teaspoon salt
freshly ground black pepper

In a 3-4 quart saucepan bring the stock to a boil. Add the chicken breast and cook for 15 minutes or until done. Remove, cut into bite sized pieces and set aside.

In another 3-4 quart saucepan, over moderate fire, heat the oil and saute the onion, mushrooms and cumin for 5 minutes stirring frequently.

Add the curry powder, cook for a minute or so and then add the flour. Stir for a few more minutes and then, very slowly, add the stock, stirring vigorously, and bring to a boil. Add the celery, tomato, salt and a few grindings of black pepper. Turn the heat to low and simmer, covered, for 30 minutes until the celery is tender. Add the reserved chicken, heat and serve in a soup tureen or bowl.

### Zucchini, *Leccinum* and Potato Soup

to serve 4

2 tablespoons vegetable oil
1 tablespoon butter
1 medium onion, chopped medium fine
2 tablespoons flour
3½ cups chicken or vegetable stock (or use 2 bouillon cubes and
    3½ cups water)
2 medium potatoes, scrubbed and cut into 1″ cubes
4 ounces sliced, frozen *Leccinum* (about 1⅓ cup)
10 ounces zucchini, fresh or frozen, chopped into bite sized pieces
½ teaspoon salt
½ teaspoon white pepper
1 cup milk

In a 3-4 quart saucepan over moderate fire, heat the oil and melt the butter. When the foam subsides, add the onion. Saute for 5 minutes until translucent and add the flour. Stir for a minute or two and add the stock. Bring to a simmer, stirring frequently. Add the potatoes, *Leccinum*, zucchini, salt and white pepper. Bring to a boil and immediately reduce the heat to low. Simmer, covered, for 30 minutes until the potatoes are tender, stirring occasionally. Add the milk and heat but do not boil. Serve.

*This is a very wintery soup, hearty and an excellent way to use up the frozen zucchini from last summer's garden. The mushroom and white pepper make it unusual.*

### Cucumber Soup, Chinese Style

to serve 4

two dried Pine Mushrooms
1 teaspoon cornstarch
1 tablespoon cold water
1 medium cucumber
3 cups chicken or vegetable stock (or 2 bouillon cubes and
   3 cups water)
2 teaspoons soy sauce
2 teaspoons pale dry sherry
1 teaspoon sesame seed oil

In a small bowl, soak the mushrooms in hot water and cover for 15 minutes or until soft. Drain, discarding the mushroom water, and slice if dried whole.

In a small cup, combine the cornstarch and one tablespoon of cold water.

Peel the cucumber. Cut in half lengthwise, scrape out the seeds with a small spoon and slice each half crosswise into 1/8 inch slices.

In a 2-3 quart saucepan, bring the stock to a boil. Add the soy sauce and wine, stir and add the cucumber and mushroom. Cook for 10 seconds (no longer, the vegetables should remain crisp), recombine the cornstarch and water mixture and add it to the soup. Mix and immediately place in a soup tureen or bowl garnished with the sesame seed oil.

*Sesame seed oil can be purchased at oriental food stores. It is not the same as that carried by health food stores.*

# Sauces, Garnishes and Pickles

**Mushroom Sauce Spiced with Cumin Seed and Coriander**
to make about 3 cups

2 tablespoons butter
1 medium onion, finely chopped
6 ounces frozen, sliced *Boletus edulis* (about 2 cups)
¼ teaspoon cumin seed
¼ teaspoon ground coriander
½ teaspoon salt
freshly ground black pepper
2 tablespoons flour
½ cup sour cream
1 cup half-and-half cream
fresh parsley

In a 1-2 quart saucepan, over low heat, melt the butter. When the foam subsides, saute the onion, mushroom and cumin seed for 5 minutes. Stir in the coriander, salt and a few grindings of black pepper. Cook 30 seconds, stirring constantly. Add the flour and, still stirring, cook for a minute or so. Add the sour cream and half-and-half. Allow to thicken but do not boil. Serve over toasted bread or cooked vegetables garnished with parsley.

### *Boletus edulis* and Sour Cream Sauce with Paprika

to make about 1 ½ cups

2 tablespoons butter
1 small clove garlic, finely chopped
1 small onion, finely chopped
3 ounces frozen, sliced *Boletus edulis* (about 1 cup)
¼ teaspoon salt
freshly ground black pepper
⅓ cup sour cream
¼ teaspoon paprika

In a small saucepan over low heat, melt the butter. When the foam subsides, saute the garlic and onion for 3 minutes until wilted. Add the mushroom and cook for 3 minutes. Add the salt and a few grindings of black pepper, mix well, and add the sour cream and paprika. Heat but don't boil. Serve over cooked vegetables.

*This is a classic French recipe.*

## Dried Puffball Garnish for Vegetables

sliced, dried puffballs
butter
garlic, finely chopped
salt
oregano
basil

Saute the puffballs in hot water for at least 15 minutes until softened. Drain, discarding the water. In a saucepan or skillet, over low heat, melt some butter. When the foam subsides add the puffball and some garlic, salt, oregano and basil. Stirring frequently, cook until lightly browned and crisp, about 15 minutes. Serve with cooked vegetables, as a garnish for soup or salad.

## Frozen Puffball and Sunflower Garnish

1½ tablespoons butter
1 ounce sliced, frozen puffballs (about ½ cup)
¼ teaspoon garlic powder
pinch sage
¼ teaspoon salt
2 tablespoons raw sunflower seeds

In a small saucepan over low heat melt the butter. When the foam subsides add the puffball, garlic powder, sage, and salt. Cook until the puffball begins to brown, about 5 minutes. Add the sunflower seed and cook until both are well browned and the puffball is crisp, about 10 minutes more. Sprinkle on cooked vegetables, soup or salad.

*These puffball recipes are often pleasing even to those who do not normally like puffballs. The crisp mushroom has the texture of the fatty part of bacon.*

## Pine Mushrooms Marinated in Soy Sauce and Port Wine

3 ounces sliced, frozen Pine Mushrooms (about 1 cup)
1 tablespoon soy sauce
1 tablespoon port wine

In a small bowl combine the frozen mushroom, soy sauce and wine. Soak for about 1 hour at room temperature.

Preheat the oven to 350°. Place the mushroom and its liquid in an overproof dish and bake uncovered for 10 minutes until the liquid is just gone. Serve as an accompaniment to plain cooked meats and vegetables.

*In the summertime, do the same to fresh Pine Mushrooms but grill over charcoal or roast over a wood fire. They are absolutely delicious!*

## Pickled Honey Mushrooms

10 cups water
5 cups cider vinegar or distilled white vinegar
   (4 to 6 percent acidity)
¾ cup pickling salt (use *only* pickling salt)
Honey Mushrooms
hot peppers
peppercorns
whole coriander seeds
basil, oregano, sage (preferably fresh)
garlic

Bring the water, vinegar and salt to a boil over high heat. Mix thoroughly.

Meantime, wash the mushrooms and blanch in boiling water. (Honey Mushrooms shrink substantially when cooked. If they are stuffed into jars when raw they shrivel up when processed in a hot water bath and the jars will be only partially full.)

Pack with selections of the other ingredients in hot, sterilized pickling jars. Pour the brine to within ½ inch of the top. Seal. To process boil for 20 minutes plus two minutes for each additional 1,000 feet above 5,000.

*Honey mushrooms take particularly well to pickling. We have tried the suggested spices and herbs in all possible combinations and they are excellent.*

# Vegetarian Dishes

## Mushrooms a la Grecque

1 cup vegetable or chicken stock
   (or 1 cup water and 1 bouillon cube)
½ cup cider vinegar
½ cup peanut oil (or olive oil)
1 teaspoon pickling salt
2 sprigs parsley
½ teaspoon basil
1 scallion, including some green, chopped
10 peppercorns
10 juniper berries
mushrooms (see below), wiped clean

Bring all the ingredients but the mushrooms to a boil. Add the mushrooms and simmer for 20 minutes. Remove the mushrooms to a glass jar or bowl. Strain the liquid, pressing on the scallion, etc. to release the flavor. Pour the liquid over the mushrooms and let stand for at least 1 hour, or refrigerate overnight.

*Most mushrooms take to pickling, but we particularly recommend the Honey Mushroom and the various Suillus.*

*Following the same recipe, these can also be put up into sterilized jars and kept for the winter months. Hot water process as with Pickled Honey Mushrooms above.*

*Olive oil is the traditional oil used for a la Grecque recipes. However, it is expensive and some don't care for its strong flavor.*

## Pear, Mushroom, Spinach with Vinaigrette

to serve 4

**the vinaigrette:**

2 tablespoons vegetable oil
1 tablespoon herb vinegar
pinch of oregano
pinch of basil
1/8 teaspoon dry mustard
salt and pepper to taste

**the spinach and pears:**

2 tablespoons peanut oil
1 medium onion, finely chopped
3 ounces frozen, sliced *Leccinum* (about 1 cup)
½ teaspoon sage
¼ teaspoon salt
freshly ground black pepper
10 ounces fresh spinach, washed and chopped
   (substitute frozen chopped)
¼ cup water
1 16-ounce can of pears in unsweetened liquid, drained

To make the vinaigrette, in a jar with a tight lid, combine the 2 tablespoons of vegetable oil, herb vinegar, oregano, basil, mustard and salt and pepper to taste. Cover and shake vigorously. Let stand at room temperature for at least 15 minutes before using.

To make the spinach, in a 12" wok or iron skillet or 3-4 quart saucepan, heat the 2 tablespoons of peanut oil over moderate heat. Add the onion and saute for 5 minutes until translucent. Add the mushroom, sage, ¼ teaspoon salt and freshly ground black pepper to taste. Stirring frequently, cook until the mushroom is defrosted and then add the spinach. Over medium heat, add the water, cover tightly and cook until the spinach is tender and the water is gone, about 7 minutes.

Place the spinach in the middle of a serving dish, surround with the canned pears and pour the vinaigrette over the top. Serve.

*This is a very pretty dish and quite festive.*

## Mushroom and Wild Green Quiche

to serve 4

### the crust: (9″ pie)

½ cup unbleached all purpose flour
½ cup whole wheat flour
2 tablespoons butter, cut in small pieces
3 tablespoons vegetable oil
¼ teaspoon salt
approximately 2 tablespoons very cold water

### the filling:

2-3 cups wild greens such as goosefoot, washed
    (about 3-4½ ounces, or substitute spinach)
2 tablespoons vegetable oil
2 large scallions, including some green, finely chopped
1 large clove garlic, finely chopped
3 ounces frozen, defrosted *Boletus edulis* or *Leccinum,*
    coarsely chopped (about 1 cup)
½ teaspoon salt
1/8 teaspoon white pepper
1 cup milk
2 eggs
½ cup grated Swiss cheese

To make the crust, in a medium sized mixing bowl mix the flours, butter, oil and salt with a fork or your fingers until they are well combined. Add the water, about a teaspoon at a time, until the dough holds its shape in a ball. Lightly flour an appropriate surface, roll out the dough into a round about 1/8″ thick and place in a buttered 9″ pie pan. Flute the edges. Refrigerate at least 1 hour before using, covered with wax paper.

If you have a quiche dish, all the better.

To make the filling, in a 2-3 quart saucepan, cook the greens for 5 minutes in rapidly boiling water until just tender. Drain thoroughly and chop fine.

In a large skillet over moderate fire, heat the oil and saute the scallion and garlic for 2-3 minutes until wilted. Add the mushroom.

Cook for another few minutes, remove from the heat and add the greens, salt and pepper. Mix well. Preheat the oven to 400°.

In a small bowl beat the milk and eggs vigorously for about 2 minutes. Stir them and the cheese into the greens mixture.

Pour the filling into the crust and bake in the middle of the oven for 45 minutes or until it is firm and browned. Remove from the oven and let stand for 5 minutes before serving.

*This crust falls apart easily and will probably have to be pieced into a pie pan. However, it is flaky and tender so well worth the effort. (From* Common Edible and Medicinal Plants of Colorado.*)*

### Polenta with Mushrooms and Swiss cheese

to serve 4

**the polenta:**

¼ cup butter
1 onion, finely chopped
¾ cup yellow cornmeal
3 cups vegetable or chicken stock
    (or 2 bouillon cubes and 3 cups water)
½ teaspoon salt
freshly ground black pepper
½ cup half-and-half cream
½ cup imported Swiss cheese, freshly grated

**the mushrooms:**

2 tablespoons butter
8 ounces frozen *Boletus edulis,* partially defrosted and chopped
    medium fine (about 2½ cups)
¼ teaspoon oregano
1 clove garlic, finely chopped
½ teaspoon salt
freshly ground black pepper

To make the polenta, in a 2-3 quart saucepan, over low heat, melt the butter. When the foam subsides, saute the onions for 3 minutes until wilted. Add the cornmeal, mix, add the stock. Turn the heat to high and bring to a boil. Reduce the heat to low, stir in ½ teaspoon salt and a few grindings of black pepper. Cook till the cornmeal starts pulling away from the sides, about 10 minutes.

Preheat the oven to 350°. Butter an approximately 13″x9″ oven-proof dish. Place half the cornmeal mixture in the bottom, pour ½ the cream over it. Spread on the other half of the cornmeal mixture and pour over it the remainder of the cream. Sprinkle on the Swiss cheese and bake uncovered, in the middle of the oven, 30 minutes until lightly browned.

Meanwhile, make the mushroom mixture. In a 12″ skillet or 1-2 quart saucepan, over low heat, melt the butter. When the foam subsides, add the mushrooms and cook for about 3 minutes.

Sprinkle on the oregano, garlic, ½ teaspoon salt and pepper, mix thoroughly. Remove from the heat and serve with the polenta.

**Eggs Stuffed with Duxelles and Mornay Sauce**

to serve 2

4 eggs

**the duxelles:**
1 tablespoon butter
1 scallion, including some green, finely chopped
4 ounces frozen *Coprinus comatus,* partially defrosted and
    chopped fine (about 4 medium sized mushrooms)
¼ teaspoon salt
freshly ground black pepper
¼ teaspoon tarragon

**the mornay sauce:**
2 tablespoons butter
2 tablespoons white flour
1 cup milk
¼ teaspoon salt
1/8 teaspoon white pepper
2 ounces Swiss cheese (about ½ cup) grated
butter

Hard boil the eggs. Peel and cut in half, separate out the egg yolk.

To make the duxelles, in a 1-2 quart saucepan or skillet, over low heat, melt the butter. When the foam subsides, add the scallion, mix and then add the mushroom, ¼ teaspoon of salt, a few grindings of black pepper and the tarragon. Cook until about 1 tablespoon of liquid remains. Remove from the heat and thoroughly mix in the cooked egg yolks. The mixture should stick together, if not add a little hot water. Stuff the egg whites with the mixture and arrange in a pan large enough to hold the whites flat but not much bigger (an 8″ pie plate is good).

To make the sauce, in a 1-2 quart saucepan over low heat, melt 2 tablespoons of butter. When the foam subsides, add the flour and cook for a minute or so. Slowly add the milk, stirring constantly until the sauce thickens. Add ¼ teaspoon salt, the white pepper and all but about 2 tablespoons of the Swiss cheese. When the cheese has melted, pour the sauce over the eggs, sprinkle on the extra cheese, dot with butter and place under the broiler until lightly browned on top.

*It has often been noted that* Coprinus *and eggs go extremely well together. This dish is no exception.*

### Artichoke with Duxelles and Bearnaise Sauce

to serve 4

**the artichokes:**

1 artichoke per person
water to cover
lemon

**the duxelles:**

4 ounces frozen *Boletus edulis* defrosted and chopped medium fine
    (about 1 cup chopped)
1 tablespoon butter
½ teaspoon peanut oil
1 scallion, including some green, finely chopped
¼ teaspoon salt
freshly ground black pepper

**the bearnaise sauce** (to make about 1 cup):

2 medium garlic cloves, coarsely chopped
1 scallion, including some green, coarsely chopped
1 tablespoon fresh parsley, coarsely chopped
½ teaspoon dried tarragon
½ teaspoon dried thyme
a small bay leaf
¼ cup tarragon vinegar
¼ teaspoon salt
1/8 teaspoon white pepper
3 egg yolks
¼ pound butter, melted

To prepare the artichokes bring the water to a boil. Trim the bottom of each so it will stand upright. With a sharp knife cut off the top inch or so of the leaves. Cut off the tip of any leaf that has a thorn with scissors. Trim and discard any brown leaves around the bottom. Rub the exposed areas with lemon to prevent discoloring.

Depending on the size, boil for 30-45 minutes or until tender when pierced with a fork. When done, remove and drain. When cool enough to handle, carefully spread the leaves back away from the choke (core) and remove it with a small spoon.

To make the duxelles, wring the chopped mushroom dry in paper towel. In a small saucepan or skillet over low fire, heat the oil and melt the butter. When the foam subsides, add the scallion. Stir and add the mushrooms, salt and a few grindings of black pepper. Cook until the liquid is gone.

To make the bearnaise sauce, in a small saucepan mix the garlic, scallion, parsley, tarragon, thyme, bay leaf, vinegar, salt and pepper. Over moderate heat bring to a boil and reduce by half. You should have about 2 tablespoons of liquid. Transfer to a small bowl, straining through a sieve with a double layer of cheese cloth over it, pressing down hard on the herbs with the back of a spoon.

Just before serving, in a small bowl beat the egg yolks vigorously for about two minutes. Beat in the vinegar mixture. Gradually add the melted butter, beating constantly. Or in a blender, simply blend the eggs and vinegar mixture together and gradually pour in the melted butter, mixing constantly.

Fill the artichokes with the duxelles, pour the sauce over the top or serve it in individual bowls for dipping.

*This sauce is not served hot. If the butter is too hot or poured in too fast, it will cook and curdle the eggs.*

## Curried *Boletus edulis* and Peas

to serve 3-4 with a traditional Indian meal, 2 Western style

1 ounce dried *Boletus edulis* (about 1 cup)
½ tablespoon peanut oil*
1 tablespoon butter*
1 medium onion, finely chopped
¼ teaspoon turmeric
½ teaspoon curry powder (garam masala if possible)
¼ teaspoon ground coriander
¼ teaspoon sage
¼ teaspoon salt
freshly ground black pepper
⅓ cup yogurt
1 cup green peas
½ cup of the mushroom water

*substitute 1½ tablespoons ghee

Soak the *Boletus edulis* in hot water for at least 15 minutes until softened. Drain, reserving the water, and slice if dried whole.

In a 12 ″ wok or iron skillet over low fire, heat the oil and melt the butter. When the foam subsides add the onion and saute for 5 minutes until translucent. Add the turmeric, curry powder, coriander, sage, salt and a few grindings of black pepper, mix well. Stirring constantly, cook for one minute. Stir in the yogurt, mix well, and add the peas, mushrooms and mushroom water. Cover tightly and simmer for 10 minutes. Serve.

*Ghee and garam masala are traditional Indian ingredients and can be purchased at specialty food stores or made at home. Ghee is essentially clarified butter that is cooked for a long time to develop a distinctive nutty flavor. Garam masala is one of many possible spice mixtures that all fall under the general name "curry powder". Indian cookbooks carry these recipes.*

*Boletus edulis is perhaps the world's favorite mushroom and there is good reason. It seems to be good anywhere, any time and does wonders to almost any dish. Here, even with a strong curry mixture, its distinctive flavor is noticeable.*

### *Leccinum* and Broccoli Chinese Style

to serve 4 at a Chinese meal, 2 Western style

1 ounce dried *Leccinum* (about 1 cup)
1 ¼ cups of the mushroom water
1 ½ tablespoons soy sauce
1 tablespoon soybean paste
1 tablespoon catsup
2 tablespoons pale dry sherry
2 tablespoons peanut oil
1 clove garlic finely chopped
2 medium stalks broccoli, peeled and cut into bite sized pieces of
   uniform size (about 3 cups)
1 teaspoon sesame oil

Soak the dried *Leccinum* in hot water for at least 15 minutes until softened. Drain, reserving the water. Slice if dried whole. In a small bowl combine the soy sauce, soybean paste, catsup and sherry.

Heat a 12 " wok or iron skillet over high fire for about 30 seconds. Add the oil and heat for another 30 seconds. Add the garlic, mix and then add the broccoli and mushroom. Stirring constantly, cook for 2 minutes. Add 1 ¼ cup of the mushroom water, cover, and still over high heat, cook for about 7 minutes until the broccoli is just tender. There should be approximately ¼ cup liquid remaining. Then add the mixture of soy sauce, soybean paste, catsup and sherry. Bring to a boil, mix well and remove to a serving platter. Sprinkle the sesame seed oil on top and serve.

*Sesame seed oil can be purchased at oriental food stores. It is not the same as that carried by most health food stores. Soybean paste also can be purchased at oriental food stores. There are dark and light soybean pastes (called miso in Japanese) and a very peppery Korean bean paste called kochee chang. They are all good but will, of course, lend different character to the dish. For optimal mushroom flavor we recommend the shiro miso or white bean paste.*

**Braised 2 Kinds of Mushrooms, Chinese Style**
to serve 4 at a Chinese meal, 2 Western style

**style no. 1:**

1 teaspoon cornstarch
1 tablespoon cold water
1 tablespoon peanut oil
1 medium clove garlic, chopped medium fine
½ teaspoon fresh ginger root, peeled and finely chopped
6 ounces sliced, frozen *Boletus edulis* (about 2 cups)
2 tablespoons pale dry sherry
2 tablespoons soy sauce

**style no. 2:**

1 teaspoon cornstarch
1 tablespoon cold water
1 tablespoon peanut oil
6 ounces sliced, frozen *Leccinum* (about 2 cups)
¼ teaspoon salt
¼ cup chicken or vegetable stock
1 tablespoon half-and-half cream

To make style number 1, in a small bowl combine the cornstarch and water. Heat a 12 " wok or iron skillet over high fire for about 30 seconds. Add the oil and heat for another 30 seconds. Add the garlic and ginger, stir for a few seconds and then add the mushroom. Cook for 3 minutes stirring constantly. Add the sherry and soy sauce, mix. Recombine the cornstarch and water and pour over the mushrooms. Mix well until the sauce thickens and place in one-half of a serving dish. Keep warm, and make:

Style number 2: in a small bowl combine the cornstarch and water. Heat a 12 " wok or iron skillet over high fire for about 30 seconds. Add the oil and heat for another 30 seconds. Add the mushroom and salt, stir fry for 1 minute and then add the chicken or vegetable stock. Still over high heat, cover and cook for 2 or 3 minutes. Recombine the cornstarch and water and pour over the mushrooms. Mix well until the sauce thickens, remove from heat, stir in the cream, and place on the other side of the serving dish.

*Traditionally, this would be served with a cooked green separating the two mushroom dishes.*

## Asparagus and Chanterelle, Chinese Style

to serve 4 at a Chinese meal, 2 Western style

⅔ cup chicken or vegetable broth
1 tablespoon soy sauce
1 tablespoon pale dry sherry
1 teaspoon cornstarch
1 tablespoon cold water
3 tablespoons peanut oil
1 cake soybean curd, cut into 1″ cubes (about 2 cups)
1 scallion, including some green, finely chopped
1 clove garlic, finely chopped
½ pound asparagus, washed, tough bottoms snapped off and
 cut into 2″ lengths
3 ounces sliced, frozen Chanterelle (about 1 cup)

In a small bowl combine the broth, soy sauce and sherry. In another small bowl combine the cornstarch and cold water.

Heat a 12″ wok or iron skillet over high fire for about 30 seconds. Add the oil and heat for another 30 seconds. Add the bean curd and brown lightly, stirring carefully but frequently. Drain on paper towels.

Still over high heat, add the scallion and garlic and stir fry 15 seconds. Add the asparagus and mushroom and stir fry until the mushrooms are separated and defrosted. Add the chicken or vegetable broth, soy sauce and sherry. Cover and still over high heat, cook for about 3 minutes until the asparagus is just tender. Mix in the bean curd. Recombine the cornstarch and water. Stirring constantly pour it over the bean curd and asparagus. Mix well until the sauce thickens, remove to a platter and serve.

*Visually, this is one of the most beautiful dishes one can imagine with the green asparagus, white bean curd and orange Chanterelle. Gastronomically, it is exquisite; the Chanterelles are outstanding here.*

## Sauteed Beancurd, Chinese Style

to serve 4 at a Chinese Meal, 2 Western style

½ ounce dried Pine Mushroom (about ½ cup)
½ cup of the mushroom water
2 tablespoons soybean paste
2 tablespoons soy sauce
2 tablespoons pale dry sherry
3 tablespoons peanut oil.
1 cake soybean curd, cut into cubes ½ "x½ "x¼ " (about 2 cups)
2 dried red hot peppers (to taste)
1 scallion, including some green, finely chopped
1 large clove garlic, finely chopped
½ teaspoon fresh ginger root, peeled and finely chopped
½ cup bamboo shoot, thinly sliced
2 stalks broccoli (cauliflower is a good substitute) peeled and
   cut into small bite sized pieces

Soak the dried Pine Mushrooms in hot water for at least 15 minutes until softened. Drain, reserving the water. Slice if dried whole.

In a small bowl combine the soybean paste, soy sauce and pale dry sherry.

Heat a 12″ wok or iron skillet over high fire for about 30 seconds. Add the oil and heat for another 30 seconds. Add the bean curd and stirring carefully but frequently, brown lightly. Drain on paper towel.

Still over high heat, add the hot pepper and stir fry until black. Add the onion, garlic and ginger and stir fry 15 seconds. Add the mushrooms, bamboo shoot, and broccoli. Stir fry for 1 minute and then pour in the ½ cup of reserved mushroom water. Cover and still over high heat, cook for about 5 minutes until the broccoli is just tender. Mix in the bean curd, soybean paste, soy sauce and pale dry sherry. Heat and serve.

### *Suillus brevipes* with Zucchini, Broccoli and Fried Noodles
to serve 4 at a Chinese meal, 2 Western style

1 ounce dried *Suillus brevipes* (about 1 cup)
2 tablespoons soy sauce
2 tablespoons pale dry sherry
2 tablespoons cornstarch
3 tablespoons peanut oil
1 large scallion, including some green, finely chopped
1 large clove garlic, finely chopped
1 teaspoon fresh ginger root, peeled and finely chopped
1 teaspoon fermented black beans, finely chopped
2 stalks broccoli, peeled and cut into bite sized pieces (substitute
  cauliflower or yellow squash)
1 cup chicken or vegetable stock
½ cup of the mushroom water
1 medium zucchini (about 6-7 ounces) washed and cut into
  strips about the same size as the broccoli
Chinese egg noodles or spaghetti, cooked and drained

Soak the *Suillus brevipes* in hot water for at least 15 minutes until softened. Drain, reserving the water. Slice if dried whole. In a small bowl combine the soy sauce, sherry and cornstarch.

Heat a 12″ wok or iron skillet over high fire for about 30 seconds. Add the oil and heat for another 30 seconds. Add the scallion, garlic, ginger root and black beans and stir fry for a few seconds. Add the broccoli and stir fry for 1 minute and then add ½ cup of stock and the mushroom water. Cover and still over high heat, cook for about 5 minutes. Add the zucchini and mushroom and cook, covered, for another 2 minutes until the zucchini and broccoli are just tender. Add the other ½ cup of stock. Recombine the cornstarch, soy sauce and sherry. Stirring constantly pour it over the broccoli and zucchini. Mix well until the sauce thickens.

Meanwhile, pour a little oil into another wok or skillet and drop in the noodles. Brown lightly on one side, flip and do the same to the other.

Pour the mushroom, broccoli and zucchini mixture over the top and serve.

Suillus *mushrooms are not a universal favorite but they grow in abundance. This dish shows them to their best advantage.*

## Meat Dishes

**Young Nettle (or Spinach), Ham and** *Suillus lakei* **Crepe**
to serve 4

**the batter:** to make 8 crepes
5 eggs
½ cup whole wheat flour
½ cup unbleached all purpose flour
1 cup milk
2 tablespoons butter, melted
¼ teaspoon salt

**bechamel sauce:**
3 tablespoons butter
3 tablespoons whole wheat flour
1½ cups milk
¼ teaspoon salt
1/8 teaspoon white pepper

**the filling:**
3 tablespoons vegetable oil
4 tablespoons shallots, finely chopped (or scallion)
1 large clove garlic, finely chopped
½ pound frozen, sliced *Suillus lakei*
10 ounces young nettle (or spinach), washed and coarsely chopped
½ teaspoon salt
freshly ground black pepper
2 tablespoons tarragon vinegar
1 cup of the bechamel sauce
8 thin slices ham
½ cup of the bechamel sauce
parmesan cheese, finely grated
chives, finely chopped

To make the crepes, thoroughly combine the ingredients listed under the batter (above) in the jar of an electric blender. Cover and refrigerate for at least one hour. (If you do not have a blender, thoroughly mix the flour, salt and egg. Add the milk and butter, mix well again, and refrigerate.)

Heat a 10-12 ″ skillet over moderate fire until water sputters instantly when sprinkled in. Spread a thin layer of butter or butter and oil combined on the bottom and sides of the pan with a pastry brush or spatula. Pour in a scant ⅓ cup of the crepe mixture, and immediately tip the pan to spread it evenly. The crepes should be about 1/8 ″ thick and 8 ″ in diameter. When the sides start to bubble, flip with a spatula. When both sides are lightly browned (about a minute on each) remove to a plate. Follow this procedure until all the batter is gone.

Make sure you regulate the heat so your crepes don't burn *or* cook too slowly.

The crepes can be made a few hours before use, covered and refrigerated.

To make the bechamel sauce, in a 1-2 quart saucepan over low heat, melt the 3 tablespoons of butter. When the foam subsides, add the 3 tablespoons of flour and cook, stirring constantly, for a minute or two. Gradually add the 1½ cups of milk, stirring vigorously and constantly, and bring to a boil. Remove from the heat and add the ¼ teaspoon salt and 1/8 teaspoon white pepper. Mix and set aside.

To make the filling, in a 10-12 ″ skillet over moderate fire, heat the oil and saute the shallots and garlic for 2-3 minutes until wilted. Add the mushrooms and cook, stirring constantly, for another 2-3 minutes. Add the nettle, ½ teaspoon salt and a few grindings of black pepper. Mix thoroughly and cook, stirring frequently until almost all the liquid is gone and the nettle is beginning to stick to the pan. Remove from the heat, add the vinegar and 1 cup of the bechamel sauce. Mix well and set aside.

The filling can be made a few hours before use, covered and set aside.

Preheat the oven to 450 °. Butter an 8 ″x12 ″ casserole or one large enough to hold the crepes (when rolled) in one layer. Spread each crepe with about 3 tablespoons of the nettle mixture, making sure the mixture is spread over most of the crepe; place a slice of ham on top, roll and place in the casserole, folded edge up. Continue in this manner until all the crepes are rolled and the nettle mixture is gone.

Spread the ½ cup of bechamel sauce on top, sprinkle with the cheese and chives, and bake in the middle of the oven for 15 minutes or until heated through.

*This dish can be combined a few hours before use, covered and refrigerated, but will require longer oven heating. (From* Common Edible and Medicinal Plants of Colorado*)*

**Veal Cutlets with Meadow Mushroom and Cream Sauce**

to serve 4

10 juniper berries (or to taste)
10 peppercorns
½ teaspoon rosemary
½ teaspoon salt
1 pound veal cutlets ¼ inch thick, pounded lightly with a kitchen mallet
whole wheat flour plus 2 tablespoons
6 ounces sliced, frozen Meadow Mushroom (about 2 cups)
2 tablespoons butter
2 tablespoons vegetable oil
1 cup half-and-half cream

With a mortar and pestle, or in a small bowl with a spoon, crush the juniper, peppercorns, rosemary and salt to a fine pulp. Firmly press this mixture into both sides of the cutlets. Dredge the cutlets in flour and shake off the excess.

In a 10-12 inch skillet, over moderate heat, melt the butter. Add the oil, heat, and saute the veal for 2-3 minutes on each side until lightly browned; do not overcook. (The meat will probably have to be cooked in batches.) Remove to a heated serving platter and cover to keep warm.

When all the veal is cooked, add the mushrooms to the skillet and, stirring constantly, saute until defrosted. Add the two table-spoons of flour, mix well and cook for a minute. Add the cream. Heat, stirring constantly, until slightly thickened, do not boil. Pour over the veal and serve.

*This receipe would also be excellent for tender cuts of wild game or pork.*

**Filet of Flounder with** *Leccinum*

to serve 4

salt
freshly ground black pepper
½ pound filet of flounder
3 ounces frozen *Leccinum,* defrosted and finely chopped (about 1 cup)
1 scallion, including some green, finely chopped
1 tablespoon fresh parsley, finely chopped
softened butter
⅓ cup dry white wine
¼ cup water
1 tablespoon butter
1 tablespoon white flour
½ cup pan drippings
¼ cup milk
½ tablespoon butter
1 teaspoon lemon juice

Lightly salt and pepper the fish. In a small bowl mix the chopped mushroom with the scallion, parsley, ¼ teaspoon salt and a few grindings of black pepper.

Thickly butter an 8″x10″ baking dish with a tight fitting cover and spread in the mushroom mixture. Arrange the fish filets on top, overlapping them slightly. Pour in the wine and water. Thickly butter one side of a sheet of waxed paper large enough to cover the top and place buttered side down on the fillets. Cover.

Preheat the oven to 350°. Bring the fish just to a boil on the stove. Bake, covered, in the middle of the oven for 10 minutes being careful not to overcook. The fish should still be springy to the touch but white, not translucent. Remove from the oven, drain off and reserve the liquid surrounding the fish (there should be about ½ cup), leaving the wax paper in place. Keep the fish warm while making the white sauce.

In a 1-2 quart saucepan, over low heat, melt 1 tablespoon of butter. Add 1 tablespoon of flour and stir for a minute or two. Add ½ cup of the pan drippings from the fish, stirring constantly.

Gradually thin out with the milk. Just before serving, stir in the ½ tablespoon of butter and the lemon juice. Pour the sauce over the fish and serve.

### Elk Stew with *Boletus edulis*

to serve 4-6

½ pound fresh side pork cut into 1½ inch lengths
2 pounds elk steak or roast, trimmed and cut into 2 inch cubes
(Most cuts of elk will work in this recipe, but the tougher ones
will require longer cooking and thus more liquid.)
rendered pork fat (from the above side pork)
12 small white pearl onions, peeled (to ease peeling drop in boiling
water for 30 seconds)
3 sprigs parsley, finely chopped
2 large scallions, including some green, finely chopped
1 large carrot, finely chopped
2 cloves garlic, finely chopped
3 tablespoons flour
2 cups beef stock
1 cup burgundy wine
1 cup water
1 tablespoon tomato paste
1 bay leaf
1/8 teaspoon thyme
1 teaspoon salt
freshly ground black pepper
12 small *Boletus edulis* (or cut up large ones) washed or wiped clean
butter
fresh parsley for garnish

In a large skillet over moderate heat, brown the side pork and
remove to paper towel. Pour off all but a thin film of the fat into a
heat-resistant container for later use.

Pat the elk meat completely dry with paper towel. Fry the elk in
the same skillet a few pieces at a time (put in enough to fill the
skillet but not so any two pieces are touching--otherwise they will
not brown properly). Remove the browned pieces to a 4-6 quart
stew pot. Continue in like manner until all the meat is browned,
adding more pork fat when necessary.

When all the meat is fried, place the peeled onions in the same
skillet and brown lightly. Remove and set aside.

Add enough pork fat to the skillet to raise the amount to 3 table-spoons. Saute the chopped parsley, scallion, carrot, and garlic for 5 minutes or until completely wilted. Stir in the flour and cook for another minute or so. Add the beef stock to the skillet, stirring constantly. Bring to a simmer stirring in all the browned bits that cling to the pan. Pour over the elk, add the wine and water, bring to a simmer over high heat, stirring constantly. Mix in the tomato paste, bay leaf, thyme, salt and a few grindings of black pepper. Add all the fried side pork.

Over low flame, simmer covered, for 2-3 hours until tender, stirring occasionally. Add more water or beef stock if necessary. Twenty minutes before serving, saute the mushrooms in a little butter. Add the mushrooms and the onions to the stew. Cook until tender. Serve garnished with parsley.

### Steak with *Leccinum* and Madeira Sauce

steak (T-bone, porterhouse, rib, filet or other tender cut)
salt
freshly ground black pepper
1 tablespoon vegetable oil
½ tablespoon butter
3 ounces sliced, frozen *Leccinum* (about 1 cup)
1 tablespoon Madeira wine

Liberally sprinkle salt and pepper on the steak and push in with the fingers. In a large skillet heat the oil over moderate heat and fry the steak until desired degree of doneness. Remove and keep warm.

In the same skillet, over low heat, melt the butter, add the *Leccinum*. Stir constantly until defrosted and they release their juice. Turn the heat to high and, stirring constantly, add the wine. Boil 30 seconds and serve over the steak.

*By changing wines or mushrooms this dish can be altered dramatically. Chanterelle, Boletus edulis, Suillus, and any number of other mushrooms work well. Ports, sherries, marsala and other sweetish wines each add their own character.*

**Sausage and** *Boletus edulis*

to serve 2

4 sausages (or the equivalent in ground sausage)
1 small onion, finely chopped
3 ounces sliced, frozen *Boletus edulis* (about 1 cup)
¼ teaspoon basil
¼ teaspoon oregano
¼ teaspoon salt
freshly ground black pepper
1 tomato peeled, seeded, and chopped medium fine

In a 12″ skillet brown the sausage over low to moderate heat. Drain on paper towel and cut into bite sized pieces.

Pour off all but 2 tablespoons of the fat remaining in the pan and over moderate heat, saute the onion for 5 minutes until translucent. Add the mushrooms, basil, oregano, salt and a few grindings of black pepper. Let the mushrooms release their juice and then add the tomato. Return the sausage to the pan, cover and over very low heat, simmer for 15 minutes. If there is not enough liquid add water to allow for this cooking time. Serve.

*A high quality sausage should be used for this dish but it is good with many varieties, from spicy Italian to a milder German. The flavor of the* Boletus edulis *comes through no matter what.*

## Roast Duckling with Mushroom Sauce

4-5 pound duck, defrosted
salt
freshly ground pepper
1 tablespoon of fat from the duck
1 tablespoon butter
1 scallion, including some green, finely chopped
6 ounces sliced, frozen *Boletus edulis* (about 2 cups)
¼ teaspoon salt
¼ cup pale dry sherry
2 tablespoons demi glaze (see below)
1 tablespoon chopped parsley

Preheat the oven to 350°. Wash and dry the duck inside and out, sprinkle salt and freshly ground black pepper inside. Prick the skin and cook in the middle of the oven 35 minutes per pound for well done duckling. Allow to sit 15 minutes before serving to let the juices settle.

To make the sauce, in a small saucepan over low fire heat the duck fat and melt the butter. When the foam subsides, add the scallion, stir and then add the mushroom, ¼ teaspoon salt and a few grindings of black pepper. Saute 2 minutes and add the sherry. Reduce the liquid by half. Add the demi glaze, mix in the parsley and pour over the duck.

*A demi glaze is a thickened brun de veau sauce made by boiling down 2 cups brun de veau into 2 tablespoons thick glaze. A quick substitute would be to boil down a mixture of 1 cup chicken broth, 1 cup beef broth and 1 tablespoon tomato puree.*

*This mushroom sauce is another famous French classic, Chasseur (Hunter's) sauce.*

*This recipe defies description. The flavor enhancing effect of mushrooms is outstanding here and the shiny sauce highlights the duck.*

### Rock Cornish Game Hens Stuffed with Wild Rice and Oyster Mushrooms

to serve 4

2 tablespoons butter
1 large clove garlic, finely chopped
1 small onion, finely chopped
¼ cup wild rice, washed thoroughly
1 ½ cups chicken or vegetable stock
3 pork sausages (about 3 ounces)
the livers of the birds
3 ounces Oyster mushrooms wiped or washed clean, sliced
   (about 1 cup)
1 tablespoon fresh parsley, finely chopped
½ teaspoon marjoram
pinch rosemary
salt
freshly ground black pepper
4 rock Cornish game hens
melted butter

In a 10-12 ″ skillet over moderate heat melt 2 tablespoons of the butter. Add the garlic and onion and saute for 10 minutes until beginning to turn gold. Stir in the washed rice and mix well.

Meanwhile, in a small saucepan bring the stock to a boil. Add the broth to the rice mixture, bring back to a boil, turn the heat to low and simmer, covered, for one hour or until the liquid is absorbed. Remove to a bowl.

In the same skillet cook and brown the sausage, remove and chop coarsely. Add the livers, brown on both sides, remove and chop fine. Pour off all but one tablespoon of fat and saute the mushrooms. Stir in the parsley, marjoram, rosemary, salt and a few grindings of black pepper. Mix in the rice mixture.

Preheat the oven to 350°. Wash the birds inside and out under cold running water. Dry thoroughly with paper towel and sprinkle the inside with salt and pepper. Just before baking, stuff the hens loosely. Bake for one hour or until done, basting every 15 minutes with melted butter.

**Shish Kebob**

to fill four skewers

1 two-pound leg of lamb
2 small-medium onions, cut in quarters from root end to stem
  and separated
2 green peppers, seeded and cut into approximately 1½ " squares
Pine Mushrooms, wiped clean
cherry tomatoes

**the marinade:**

1 large clove garlic, finely chopped
1 teaspoon fresh ginger root, peeled and finely chopped
½ cup vegetable oil (olive oil)
juice of one lemon, strained
1 cup dry white wine
1 tablespoon coriander leaves
1 teaspoon ground cumin
1 teaspoon turmeric
½ teaspoon salt
freshly ground black pepper

Remove all fat and gristle from the lamb and cut into approximately 1 " cubes.

Combine all the ingredients for the marinade. (If the Shish Kebob is to be cooked outside, any stainless steel or glass dish large enough to hold all the ingredients will do. However, if it is to be broiled under an oven broiler, the best marinade pan is an 8 "x12 " pyrex baking dish or any similarly shaped dish.)

Put the meat, green pepper, onion and mushroom into the marinade and mix well. Let sit at room temperature for at least 2 hours but not more than 4, stirring occasionally; or refrigerate overnight. Add tomatoes to the marinade, mix, and string on skewers, alternating vegetables and meat.

If cooking outdoors, cook over hot coals, turning and basting occasionally (or cook over a campfire).

If cooking under a broiler, preheat. Leave the marinade juices in the pan, place the skewers across the top and broil for approximately 15 minutes turning and basting occasionally.

**Sweet Sour Pork, Chinese Style**

to serve 4 at a Chinese meal, 2 Western style

12 ounces pork tenderloin cut into 1 "x1 " squares (about 4 big chops
 after being boned and trimmed)
1 tablespoon soy sauce
2 teaspoons cornstarch
2 tablespoons catsup
½ tablespoon honey
2 tablespoons soy sauce
2 tablespoons pale dry sherry
½ cup water
1 teaspoon cornstarch
1 tablespoon cold water
peanut oil
1 large clove garlic, finely chopped
½ teaspoon fresh ginger root, peeled and finely chopped
6 ounces sliced, frozen *Boletus edulis* (about 2 cups)
1 medium onion, chopped medium fine
1 small carrot peeled, and sliced into rounds
½ cup green pepper washed, and cut in cubes about the same
 size as the carrot
1 tablespoon rice vinegar (or substitute white wine vinegar)
4 slices of pineapple, cut about the same size as the carrot

In a glass or stainless steel bowl, combine the pork with 1 tablespoon of soy sauce and 2 teaspoons of cornstarch. Let stand at
room temperature for 15-30 minutes (no longer).

In a small bowl combine the catsup, honey, 2 tablespoons soy
sauce, 2 tablespoons pale dry sherry and ½ cup water. In another
combine the 1 teaspoon cornstarch and 1 tablespoon cold water.

In a 12 " wok over high fire, heat approximately 4 cups of peanut
oil to 360°. Fry the pork squares, separating them as much as possible, until they are a rich golden brown. Remove to paper towels.

Remove all but 2 tablespoons of the oil. Still over high heat, add
the garlic and ginger root. Stir fry for a few seconds and then add
the mushroom, onion, carrot, and green pepper. Stir fry for 2
minutes and add the mixture of catsup, honey, soy sauce, sherry

and ½ cup water. Cover and still over high heat, cook for about 3 minutes until the carrot is just tender. Add the fried pork, vinegar and pineapple. Bring to a boil. Recombine the cornstarch and water and pour over the pork. Mix well until the sauce thickens and serve.

**Shrimp in Spicy Sauce, Chinese Style**

to serve 4 at a Chinese meal, 2 Western style

½ ounce dried *Leccinum* (about ½ cup)
½ cup of the mushroom water
12 medium shrimp
2 tablespoons catsup
1 tablespoon soy sauce
1 teaspoon cornstarch
1 tablespoon cold water
2 tablespoons peanut oil
2 dried red hot peppers (to taste)
1 scallion, including some green, finely chopped
½ teaspoon fresh ginger root, peeled and finely chopped
1 cup peas

Soak the dried *Leccinum* in hot water for at least 15 minutes until softened. Drain, reserving the water. Slice if dried whole. Shell and devein the shrimp.

In a small bowl combine the mushroom water, catsup and soy sauce. In another small bowl combine the cornstarch and water.

Heat a 12 " wok or iron skillet over high fire for about 30 seconds. Add the oil and heat for another 30 seconds. Add the hot pepper and stir fry until black. Stirring constantly, add the scallion and ginger and cook for a few seconds, then add the peas, mushroom and shrimp and stir fry for 1 minute. Add the mushroom water, catsup and soy sauce. Cover and still over high heat, cook for about 2 minutes until the shrimp is pink and peas are just tender.

Recombine the cornstarch and water. Stirring constantly pour it over the shrimp. Mix well until the sauce thickens. Remove to a platter and serve.

*Here the unusual color of* Leccinum *is featured, contrasting with the green peas, pink shrimp and red sauce. Its chewy texture is also complementary.*

## Dry Shredded Beef, Chinese Style

to serve 4 at a Chinese meal, 2 Western style

8 ounces flank steak, cut into thin strips ¼ " wide and 2 " long
2 tablespoons soy sauce
2 teaspoons pale dry sherry
½ ounce dried *Leccinum* (about ½ cup)
4 tablespoons peanut oil
½ teaspoon hot red pepper flakes (to taste)
½ teaspoon fresh ginger root, peeled and finely chopped
1 large carrot, peeled and cut into strips 1/8 " wide and 2 " long
2 stalks celery, washed and cut into strips 1/8 " wide and 2 "long

Place the flank steak in a glass or stainless steel bowl and thoroughly mix with the soy sauce and sherry. Let stand at room temperature for 15-30 minutes (no longer).

Soak the dried *Leccinum* in hot water for at least 15 minutes until softened. Drain, discarding the water. Slice into thin strips 1/8 " wide and 2 " long.

Heat a 12 " wok or iron skillet over high fire for about 30 seconds. Add the oil and heat for another 30 seconds. Add the beef and stir fry until very dark, dry, and crisp. This will take about 10 minutes. Remove to a bowl.

Discard all but 2 tablespoons of the oil and, still over high heat, add the hot peppers and ginger. Cook 15 seconds and add the mushrooms, carrots and celery. Stir fry for a minute or so; the vegetables should stay quite crisp. Add the beef, mix and serve.

*This recipe can also be made with mushrooms replacing the beef. Eliminate the dried* Leccinum *and use fresh or frozen* Leccinum *or* Boletus edulis *instead of the beef. Do not marinate, but fry in the same manner as the beef until the mushroom is crisp and golden. Then pour in the 2 tablespoons of soy sauce and 2 teaspoons of sherry, stir and remove from the heat. Continue with the rest of the recipe.*

## Chicken with *Leccinum* and Peanuts, Chinese Style

to serve 4 at a Chinese meal, 2 Western style

1 boned chicken breast, skin removed and cut into 1 " cubes
2 teaspoons pale dry sherry
2 teaspoons cornstarch
2 teaspoons soy sauce
2 tablespoons soy sauce
2 tablespoons pale dry sherry
1 teaspoon sesame seed oil
2 teaspoons cornstarch
½ teaspoon vinegar (rice is best but white wine will do)
peanut oil
1 dried red hot pepper (to taste)
1 teaspoon fresh ginger root, peeled and finely chopped
3 ounces sliced, frozen *Leccinum* (about 1 cup)
2 ounces Chinese pea pods, destringed and cut into thin strips
    lengthwise
4 tablespoons salted peanuts

In a glass or stainless steel bowl, combine the chicken with 2 teaspoons of sherry, 2 teaspoons of cornstarch and 2 teaspoons of soy sauce. Let stand at room temperature for 15-30 minutes (no longer).

In a small bowl combine the 2 tablespoons of soy sauce, 2 tablespoons of sherry, 1 teaspoon of sesame seed oil, 2 teaspoons of cornstarch and ½ teaspoon vinegar.

In a 12 " wok over high fire, heat approximately 4 cups of peanut oil to 360°. Fry the chicken cubes, separating them as much as possible, until they are a rich golden brown. Remove to paper towels.

Remove all but 2 tablespoons of the oil. Still over high heat, add the red pepper and fry until black. Add the ginger, mushroom and pea pods and stir fry for 1 minute. Return the chicken, add the nuts. Recombine the soy sauce, sherry, sesame oil, cornstarch and vinegar mixture and pour over the chicken. Mix well until the sauce thickens and serve.

*Sesame seed oil can be purchased at oriental food stores. It is not the same as that carried by health food stores.*

*Although we have taken some liberties, this recipe and the beef and shrimp recipe that precede it are all peppery Chinese classics. The dark, unusually textured, tasty* Leccinum *is particularly attractive here. Oyster Mushroom and Pine Mushroom would also be excellent.*

*To make a very different, but good, sort of dish, the red pepper can be omitted.*

### Cantharellus cibarius

Why, the very name
sings.
The r, the l's
make of any tongue
a small bell.

Colored aspen leaves
dot
lichen-grey rock;
juniper, pine-duff
underfoot.

Suddenly orange
pipes
raise the litter.
Wrinkled, coppery:
chanterelles--

the telltale fragrance
like
a peach orchard.
Our minds skitter through
the woodland.

And are we to be
bound
by the forest
circle home to it
forever

like a fairy tale?
Yes.
Talk and laughter
are some of its own
melody.